SpringerBriefs in Computer Science

SpringerBriefs present concise summaries of cutting-edge research and practical applications across a wide spectrum of fields. Featuring compact volumes of 50 to 125 pages, the series covers a range of content from professional to academic.

Typical topics might include:

- A timely report of state-of-the art analytical techniques
- A bridge between new research results, as published in journal articles, and a contextual literature review
- A snapshot of a hot or emerging topic
- An in-depth case study or clinical example
- A presentation of core concepts that students must understand in order to make independent contributions

Briefs allow authors to present their ideas and readers to absorb them with minimal time investment. Briefs will be published as part of Springer's eBook collection, with millions of users worldwide. In addition, Briefs will be available for individual print and electronic purchase. Briefs are characterized by fast, global electronic dissemination, standard publishing contracts, easy-to-use manuscript preparation and formatting guidelines, and expedited production schedules. We aim for publication 8–12 weeks after acceptance. Both solicited and unsolicited manuscripts are considered for publication in this series.

More information about this series at http://www.springer.com/series/10028

Hassan Ugail · Ahmad Ali Asad Aldahoud

Computational Techniques for Human Smile Analysis

 Springer

Hassan Ugail
Faculty of Engineering and Informatics
University of Bradford
Bradford, West Yorkshire, UK

Ahmad Ali Asad Aldahoud
Faculty of Engineering and Mathematics
University of Bradford
Bradford, West Yorkshire, UK

ISSN 2191-5768   ISSN 2191-5776 (electronic)
SpringerBriefs in Computer Science
ISBN 978-3-030-15380-9   ISBN 978-3-030-15381-6   (eBook)
https://doi.org/10.1007/978-3-030-15381-6

This Springer imprint is published by the registered company Springer Nature Switzerland AG
The registered company address is: Gewerbestrasse 11, 6330 Cham, Switzerland

To all those who smile.

Preface

It is often said that the face is a window to the soul. Bearing a metaphor of this nature in mind, one might find it intriguing to understand, if any, how the physical, behavioural as well as emotional characteristics of a person could be decoded from the face itself. With the increasing deductive power of machine learning techniques, it is becoming plausible to address such questions through the development of appropriate computational frameworks. Though there are as many as over twenty-five categories of emotions one could express, regardless of the ethnicity, gender or social class, across humanity, there exist six common emotions namely happiness, sadness, surprise, fear, anger and disgust—all of which can be inferred from facial expressions. Of these facial expressions, the smile is the most prominent in social interactions.

The smile bears important ramifications with beliefs such as it makes one more attractive, less stressful in upsetting situations and employers tending to promote people who smile often. Even pockets of scientific research appear to be forthcoming to validate such beliefs and claims, e.g. the smile intensity observed in photographs positively correlates with longevity, the ability to win a fight and whether a couple would stay married. Thus, it appears that many important personality traits are encoded in the smile itself. Therefore, the deployment of computer based algorithms for studying the human smiles in greater detail is a plausible avenue for which we have dedicated the discussions in this book.

In this book, we discuss the recent developments in computational techniques for automated non-invasive facial emotion detection and analysis with particular focus on the smile. By way of application, we discuss how genuine and non-genuine smiles can be inferred, how gender is encoded in a smile and how it is possible to use the dynamics of a smile itself as a biometric feature.

Bradford, UK
February 2019

Hassan Ugail
Ahmad Ali Asad Aldahoud

Acknowledgements

This book is a result of years of research into the topic of computer based human facial analysis by the authors. To this end, we are grateful for the research support and funding we have received from a variety of sources. Notably, we would like to acknowledge the doctoral scholarship provided for Ahmad Al-dahoud by Al-Zaytoonah University of Jordan. We gratefully acknowledge funding we have received from the European Union's Horizon 2020 Programme H2020-MSCA-RISE-2017, under the project PDE-GIR with grant number 778035 and also funding from the University of Bradford's Higher Education and Innovation Fund. Undoubtedly, it takes a fair amount of dedication and continuous effort to put together a book—usually work on a book falls outside the regular lines of the day to day academic tasks. As a result, a support base outside the immediate academic environment is essential to complete a book of this nature. We feel blessed to have received such support from our dear family members, our friends, our colleagues and our collaborators. Without their support and continued encouragement, this book would not have seen the light of the day. Our heartfelt thanks, therefore, goes to all of them. Thank you!

Contents

Acronyms

AAM	Active Appearance Model
ASM	Active Shape Model
AU	Action Unit
EMFACS	Emotional Facial Action Coding System
EMG	Electromyography
FNP	Facial Normalisation Parameter
GUI	Graphical User Interface
k-NN	k-Nearest Neighbour
LBP	Local Binary Patterns
SIFT	Scale Invariant Feature Transform
SVM	Support Vector Machine

Chapter 1
Facial Emotional Expressions

How many times have you smiled today? How many times have you frowned today? Ever thought of being in a state of self-consciousness to be able to relate your own mood with your facial emotional expressions? Perhaps with our present-day busy lives, we may not consider these as crucial questions. However, as researchers uncover more and more about the human emotional landscape they are learning the importance of understanding our emotions. They are also learning the importance of being able to accurately measure our emotional status. And with the advent of ever increasingly powerful automated computational tools what better way is there to understand our emotions than non-invasively through our face itself?

As the saying goes "your smile is the window to your soul", facial emotional expressions provide us with an interface that connects our physical body with our mind. In other words, facial emotional expressions provide a window through which we may peek into the mind of someone. In this sense, understanding our emotional expressions can help us better infer our feelings, our reactions, our level of awareness and our level of connectively with a given environment.

For example, people who are unwell are generally likely to have a sad face. In fact, there exist quite a number of studies that make strong links with our emotional expressions and the salient social behavioural cues. The work of Seidel et al. [13] clearly highlights this. They specifically studied the impact of facial emotional expressions on behavioural tendencies. They utilised photos of evoked facial expressions of four basic emotions namely happiness, sadness, anger and disgust. They tried to associate these emotions with two very opposite human behaviour, i.e. approach—the action of focussing or advancing closer and avoidance—the action of keeping away from or not doing something. Results of their studies indicate that happiness is strongly associated with approach while anger is associated with avoidance. Similarly, sadness invokes conscious avoidance and disgust appeals to a clear behaviour of avoidance too.

Of all facial emotional expressions, the smile expression is the most interesting. The smile expression is what occurs most commonly on our face as we go about our daily activities. The smile also appears to be the most accurately measurable facial emotional expression which can be mapped with activities of physical behaviour or

H. Ugail and A. A. A. Aldahoud, *Computational Techniques for Human Smile Analysis*, SpringerBriefs in Computer Science, https://doi.org/10.1007/978-3-030-15381-6_1

personality traits. On the face of it, a smile may appear to be a relatively simple human act. However, if we try figuring out the procedures and processes involved in a smile, we may realise that this activity is far from simple.

You hear a pleasant pun, or you witness a funny image, or you converse with someone dear to you and these activities may feed sensory information to very specific regions of your brain. That results in the brain invoking very specific actions on the facial muscles. In the case of the emotion invoking a true state of happiness, two major facial muscles get aroused. They are the zygomatic major muscle—which resides in the cheek—and the orbicularis oculi—which encircles the eye socket. The results are visually profound with the lip corners clearly curling upwards while the outer corners of the eyes showing pronounced wrinkles depicting the feet of a crow. This entire sequence of action is short lived and typically will last around two thirds of a second. And depending on the nature and the frequency of the smiles we bear on the face, they can be associated with profound behavioural cues or personality traits.

Your smile could predict how long you would live. As bizarre as it may sound, the work of Abel and Kruger [1] demonstrate this may be the case. In their work, they classified photographs of 230 major league baseball players who played during the 1950s. They looked at the amount of smile on the players' faces, particularly the amount of genuine smiles. By analysing the smile data of 184 deceased players from the sample dataset, they concluded that around 70% of people who had prominent smiles on their faces lived to the age of 80. On the other hand, half of the people who did not express smiles in their photos did not reach that age. Thus, after all, it appears, there is a good reason to smile more, should we have the desire to live longer.

A smile may be able to predict the success of a marriage. In the work of Hertenstein et al. [9], they trolled through people's college yearbook photos to identify how much people were smiling in them. They measured the smiles based on the amount of curl on the mouth and the amount of crow's feet present around the eyes. They then rated the smiles on the photos from 1 to 10 according to their intensity. Interestingly, none of the couples within the top 10% of higher smile intensity had divorced. Whilst the researchers cannot pinpoint to the exact reason for this peculiar correlation, they believe it may be due to the positive attitude people posses which may hold their marriage together, or it may be happy people have a tendency to attract other happy people, or the amount of smile on your face ensures a level of compromise which is an essential criterion for a happy and successful marriage.

Should you wish to win a physical context, smiling just before it may not be such a good idea. The work of Kraus and Chen [11] indicate that smiles may be associated with decreased physical dominance. In their study, they used pre-fight photographs of fighters to see the amount of smiles on their faces. They found that those who smiled more before the match performed more poorly during the fight in relation to their less intensely smiled counterparts. They extended their experiment by taking photos of fighters and digitally manipulating them to show more smiles

Fig. 1.1 An impression of Duchenne's stimulation of various facial muscles with electrical current to invoke emotions

on their faces. These were then shown to untrained observers who were asked to judge the ability of the fighters to win a fight. The results indicate that the untrained observers judge those who smile more to be less hostile and less aggressive thus marking them down in terms of winning a fight, in comparison with the counterparts who smiled less.

As we discuss the smiles within a scientific context, the term Duchenne which relates to genuine smiles requires specific emphasis here. This term is due to the French neurologist Guillaume-Benjamin Duchenne who stimulated facial muscles through the application of direct electrical current to the face to purposefully invoke certain facial expressions during the 1860s. Figure 1.1 illustrates an impression of Duchenne's stimulation of various facial muscles with electrical current to invoke emotions. While such invasive techniques of human experimentation are seriously questionable from the point of view ethical considerations today, Duchenne at the time managed to get away with it.

Duchenne postulated that the face is some kind of a map which can be utilised to understand the mental states of a human. And in his own words, he said that by reading facial emotion expressions it may help reveal "the accurate reading of the soul's emotions". Through his numerous experimentation on the human face, he was one of those who was able to map specific facial emotions with specific muscle groups. The fundamental theories he put forward in 1862 on the relationship between facial emotions and facial muscle movement still holds valid [7].

Thus, it was Duchenne who set the stage for the scientists of the 20th and 21st century to undertake detailed studies of human behaviour and human traits through the "window to the soul" of facial emotional expressions.

1.1 Measuring Facial Emotional Expressions

Emotional expressions on the face display the deeper cognitive processes arising from the functionality of the brain. It is essential for processing social interactions and is crucial to get on with our daily lives. From an evolutionary point of view, being able to read and understand facial expressions are vital for survival. Thus, the ability to accurately and efficiently process facial emotional expressions is fundamental to many interpersonal interactions. Beyond this fundamental need, research appears to be uncovering the importance of understanding emotional expressions which also have implications from the point of clinical deficits where the inability to process certain emotional expressions can be attributed to specific disorders.

Thus, there is no doubt that there exist a wide range of applications of facial emotional expressions analysis. These include better social interactions, assistance with various psychological and social studies, understanding and interpretation of clinically important personality disorders, a better understanding of personality traits and behavioural science and designing of better avatars for human-machine interfaces.

We all do experience emotions, but each one of us perhaps experiences them differently. Not all experts do agree on the criteria for measuring or studying emotions because emotions are complex and have both physical and non-physical components. However, all researchers agree that emotions are composed of subjective feelings, resulting in physiological responses and expressive behaviour. A subjective feeling can be one that is defined as an individual experience due to a certain emotion. Thus, the subjective feeling is very hard to measure or describe. In this sense, such definition of emotion or feeling may not have a universal adhering to it. For example, when it comes to feeling anger, two people experiencing the emotion may not experience or describe it in the same way.

Early investigations into the facial emotional expressions raised the important question of whether facial emotional expressions are universal across humanity regardless of ethnicity or culture. Among the various theories of emotions which have emerged over the past 50 years or so the neurocultural theory of emotions stands tall to date. This is due to the pioneering work of Charles Darwin who carried out systemic experiments on facial emotions in a cross cultural setting [4]. Darwin not only studied ordinary people across cultures but he also looked at the facial expressions of young children, people born blind and even people with mental illnesses. From his extensive experimentations, Darwin was able to draw strong similarities and universalities in facial emotions across ethnicities and cultures.

Following from Darwin's pioneering work, Ekman postulated the universalities of facial emotions across cultures with six basic emotions namely happiness, sadness, surprise, anger, disgust and fear. Ekman later extended his emotion model to include 11 additional emotions which can be inferred from the face namely, excitement, amusement, embarrassment, pride in achievement, shame, guilt, contempt, contentment, relief, satisfaction and sensory pleasure. However, considering Ekman's basic and extended emotions, the emotion of feeling happiness related to the smile expression can only be observably related to the underlying physiological state. The rest of the expressions face significant controversial challenges in term of their acceptability within the scientific community.

As far as methods for measuring facial emotion expressions are concerned, at present, there are three of them. They are the Facial Action Coding system (FACS)—based on the observed measurements of individual facial muscle movements, Electromyography (EMG)—based on measuring the facial muscle activity through amplified electrical impulses that are generated by the underlying muscle fibres and computational approaches where image and video signal analysis techniques are employed to infer the underlying facial emotions.

1.1.1   Facial Action Coding System

FACS is a system used to categorise human facial muscle movements based on their appearance on the face. It is claimed that FACS was originally developed by a Swedish anatomist named Carl-Herman Hjortsjo, though the method is widely attributed to Paul Ekman who worked extensively on it during the late 1970s. According to Ekmans work, FACS is a representation of facial expressions constructed by muscles or a group of muscle movements. These movements are called "Action Units" (AUs). By observing AUs, FACS can detect and measure facial expressions. Ekman has identified 46 AUs, where a single AU or group of AUs can be used to represent facial muscles. For instance, AU1 corresponds to the raising of inner eyebrows while AU27 corresponds to the stretching of the mouth. This technique can help humans identify emotions using video analysis of a subject's face to document the facial expression changes through the representation via AUs.

FACS can be an unbiased mechanism to thoroughly study the facial emotions. However, FACS coding appears to be cumbersome and laborious. One would need hours of special training to be a proficient FACS coder. Ekmans FACS manual is over 500 pages. As a result, only a small number of FACS coders can successfully code facial expressions using FACS and therefore there exist limited studies based on FACS coding.

In 1981, Ekman presented the EMFACS (Emotional Facial Action Coding System). According to Ekman, EMFACS is more economical in comparison to his original FACS as coders are not required to detect each muscle change; instead, they decide if a group of muscle changes is associated with specific emotions. Accordingly, EMFACS is a shorter version of FACS that takes less time to analyse and produces the same results. Table 1.1 shows some common facial emotions and related AU groups.

Table 1.1 Examples of some basic facial emotions and the corresponding EMFACS coding

Facial emotion	EMFACS action units
Happiness	AU6 + AU12
Sadness	AU1 + AU4 + AU15
Fear	AU1 + AU2 + AU4 + AU5 + AU7 + AU20 + AU26
Anger	AU4 + AU5 + AU7 + AU23
Disgust	AU9 + AU15 + AU16

1.1.2   Electromyography

Electromyography (EMG) measures the electrical potential generated by muscle fibres as they contract and as they return to their resting positions. EMG mainly focuses on the study of the corrugator muscle (which lowers the eyebrow) and the zygomatic muscle.

For example, Dimberg and Lundquist used EMG for measuring angry and happy expressions through by monitoring the corrugator and zygomatic muscle activity [6]. Their results show that angry faces display increased corrugator muscle activity whereas happy faces displayed increased zygomatic muscle activity. Additionally, these effects were more noticeable for females, mostly for the reaction to happy faces which conclude that females are more facially reactive than males.

The advantages of an EMG system include its ability to detect minute facial muscle activities which may be useful to study and produce empiric verification on facial emotional theories. EMG also has disadvantages. These include its relative complexity, inflexibility as well as lack of non-invasiveness which greatly reduces its use in social settings.

1.1.3   Computational Techniques

The task of a computational technique is to analyse an image or a video feed from which the facial expressions are determined and classified. This is indeed a very challenging endeavour and even to date, researchers are battling with it. There are many issues that arise when one is concerned with developing an automated system to read and analyse faces. First, the pose of the face may change relative to the camera location, or the face may be seen on the camera at an angle. Robust image pre-processing techniques are required to ensure the input feed from the camera are translation, rotation and scale invariant. Second, the face may be occluded where only parts of the face are available which would make it extremely difficult to infer the definition of full face and the expression on it. Thirdly, variation in lighting may also be problematic as illumination affects the accuracy because most of the image processing algorithms work on pre-set parameters and conditions.

In general, any computational technique or model for facial expression processing has three steps. They are face detection, facial feature extraction and facial expression classification. Many methods in the past have been proposed for these tasks and we highlight some of these below.

1.1.3.1   Face Detection

Face detection is the process of identifying the face in an image and extracting the region of the image that only contains the face. This process usually involves segmentation, extraction, and verification of the face from an uncontrolled background. The most commonly used face detector for automatic facial expression analysis is the so called Viola Jones algorithm [14] and related techniques, e.g. [2]. This algorithm makes use of a cascade of filters, which are trained by Ada Boost. Here we provide some of the common methods available for the task of face detection.

Feature invariant methods: Facial features including texture and skin colour are detected using methods for edge detection, shape feature detection and Gaussian mixtures.

Template matching methods: Predefined face templates are matched using deformable models, shape templates or using active shape models.

Appearance based methods: Many methods including Eigen decomposition, clustering, Gaussian distributions, neural networks and multilayer perceptrons with classification techniques such as SVMs, polynomial kernels, Naive Bayes classifiers and hidden Markov models are utilised.

1.1.3.2   Facial Feature Extraction

Once the face is successfully detected, the next step is to extract the relevant facial features from which the expression can be inferred. The methods available for facial feature extraction can generally be divided into two categories. They are geometry based feature extraction and appearance based feature extraction.

Geometry based feature extraction: These techniques mainly look for prominent features such as the mouth, eyes and the forehead and their relative sizes as well as shapes. Aside from the shape itself, geometric techniques can be used to measure the dynamics of key facial features.

Appearance based feature extraction: These techniques look for the changes in the facial texture and shape such as the appearance and disappearance of wrinkles and other changes around key facial features. A wide range of image filters can be utilised for this process. They may include principal component analysis, independent component analysis, linear discriminant analysis, Gabor wavelets, local binary pattern analysis and optical flow.

1.1.3.3   Facial Expression Classification

The final step in a computational facial emotion analysis system is the classification of a given expression or set of expressions into an emotion. Thus, the classifier should provide the user with a meaningful interpretation of the expression, e.g. "85% happy". Many techniques are available to undertake such classification and interpretation. The general line followed by the computational models is to either use some sort of template based matching or a neural network based classification. Most of the computational methods discussed above are available at the disposal of the user. And in the next Chapter, we present a framework we have devised for automated analysis of facial expressions.

1.2   Datasets

Data is a crucial ingredient in computational research. Algorithms require data to identify and adjust the various parameters associated with, as well as to test and verify their outputs. There exist a handful of datasets which can be utilised to undertake smile related facial emotion analysis research. Here we provide brief details of some of them.

The CK+ Dataset: The Cohn-Kanade AU-Coded Facial Expression Database contains two versions: CK and CK+. CK includes 486 video sequences of facial expressions from 97 posers. Each sequence begins with a natural expression and proceeds to a peak expression. The peak of the facial expression for each sequence is FACS coded and emotion labelled. CK+ includes both spontaneous as well as posed facial expressions [12].

The MUG Dataset: This consists of image sequences of 86 subjects performing a non-posed facial expression (laboratory induced expressions). The environment was controlled where subjects were sitting on a chair in front of one camera. They sat behind a blue screen background and had a fixed light source on their face. The MUG database consists of 86 participants (35 women, 51 men), all with the Caucasian origin, between 20 and 35 years of age. The MUG dataset is not FACS coded [3].

UvA-NEMO Smile Dataset5: This dataset consists of video sequences of 400 subjects (185 females, 215 males) of spontaneous smiles of frontal faces under reasonably constant lighting conditions. To elicit spontaneous smiles, each subject was shown a short funny video segment. The smile captured starts and ends with the neutral expression. The pose of the subjects is frontal, camera-to-subject and illumination are reasonably constant across subjects [5].

Indian Spontaneous Expression Database (ISED): ISED is a spontaneous expression dataset containing near frontal face video of 50 participants recorded while watching emotional video clips. The dataset comprises of spontaneous emotions at high resolution and frame rates along with information regarding the gender of the participants. Emotions in this dataset include happiness, disgust, sadness and surprise [10].

Additionally, in [15] one would find a good survey description of some of the datasets available for research in this area.

Note, to test the methodologies we have discussed in this book, we have predominantly used the CK+ and MUG datasets.

References

1. Abel, E., Kruger, M.: Smile intensity in photographs predicts longevity. Psychol. Sci. **21**, 542–544 (2010)
2. Al-dahoud, A., Ugail, H.: A Method for location based search for enhancing facial feature detection. In: Angelov, P., Gegov, A., Jayne, C., Shen, Q. (eds.) Advances in Computational Intelligence Systems. Springer, pp. 421–432 (2016)
3. Aifanti, N., Papachristou, C., Delopoulos A.: The MUG facial expression database. In: The 11th International Workshop on Image Analysis for Multimedia Interactive Services, WIAMIS 2010, pp. 1–4. IEEE (2010)
4. Darwin, C.: The Expression of the Emotions in Man and Animals. John Murray, London (1872)
5. Dibeklioglu, H., Salah, A.A., Gevers, T.: Are you really smiling at me? Spontaneous versus posed enjoyment smiles. In: European Conference on Computer Vision (ECCV), pp. 525–538. Springer (2012)
6. Dimberg, U., Lundquist, L.-O.: Gender differences in facial reactions to facial expressions. Biol. Psychol. **30**(2), 151–159 (1990)
7. Duchenne, G.-B., Cuthbertson, A.R.: The Mechanism of Human Facial Expression. Cambridge University Press (1990)
8. Ekman, P.: Emotional and conversational nonverbal signs. In: Messing, L.S., Campbell, R., (eds.), Gesture, Speech, and Sign, pp. 45–55. Oxford University Press (1999)
9. Hertenstein, M.J., Hansel, C.A., Butts, A.M., Hile, S.: Smile intensity in photographs predicts divorce later in life. Motivation Emot. **33**(2), 99–105 (2009)
10. Indian Spontaneous Expression Database (2018). https://sites.google.com/site/iseddatabase/
11. Kraus, M.W., Chen, T.W.: A winning smile? Smile intensity, physical dominance, and fighter performance. Emotion **13**(2), 270–279 (2013)
12. Lucey, P., Cohn. J.F., Kanade, T., Saragih, J., Ambadar, Z., Matthews, I.: The extended Cohn-Kanade dataset (CK+): a complete dataset for action unit and emotion-specified expression. In: IEEE Computer Society Conference on Computer Vision and Pattern Recognition Workshops (CVPRW), pp. 94–101. IEEE (2010)
13. Seidel, E.-M., Habel, U., Kirschner, M., Gur, R.C., Derntl, B.: The impact of facial emotional expressions on behavioral tendencies in females and males. J. Exp. Psychol. Hum. Percept. Perform. **36**(2), 500–507 (2010)
14. Viola, P., Jones, M.: Robust real time face detection. J. Comput. Vis. **57**, 137–154 (2004)
15. Zeng, Z., Pantic, M., Roisman, G.I., Huang, T.S.: A survey of affect recognition methods: audio, visual, and spontaneous expressions. Pattern Anal. Mach. Intell. **31**(1), 31–58 (2009)

Chapter 2
A Computational Framework for Measuring the Facial Emotional Expressions

Abstract The purpose of this chapter is to discuss and present a computational framework for detecting and analysing facial expressions efficiently. The approach here is to identify the face and estimate regions of facial features of interest using the optical flow algorithm. Once the regions and their dynamics are computed a rule based system can be utilised for classification. Using this framework, we show how it is possible to accurately identify and classify facial expressions to match with FACS coding and to infer the underlying basic emotions in real time.

Keywords Location based search · Facial feature detection · Regions of interest · Viola Jones algorithm

2.1 Introduction

The aim of a computational facial emotional expression recognition model is to mimic humans and to possibly go beyond the face processing abilities by humans. This is, of course, a non-trivial task. Research over the last couple of decades has made a concentrated effort to achieve this, and we have today come far, e.g. [14, 15]. There exist techniques which can accurately detect the face from a video clip and process it to infer the emotions. The bulk of these techniques are based on dimensionality reduction approaches using image processing and machine learning.

In this chapter, we discuss a methodology we have adopted to develop a real-time emotion recognition engine to process video facial images. The platform we have developed is lightweight and computationally efficient. It uses standard face detection techniques combined with optical flow to infer the facial feature motions from which we can deduce the corresponding emotions.

2.2   Methodology

Our methodology has three essential components. They are the face detection, the analysis of the detected face for facial features and classification of the facial features to infer the relevant AUs or facial expressions from which we can ultimately deduce the relevant emotions. Figure 2.1 shows our proposed framework.

2.2.1   Face Detection

The problem of face detection from video images has been approached in many ways. Many methods and algorithms have been proposed to address this issue, all of them with their own strengths and weaknesses. For example, in Singh et al. [11] simply have used the skin colour to detect the face. Their method starts with the assumption that only faces appear in the image. In Pai et al. [8], enhance the work presented in [11] and add a "low pass filter" to enhance the detection of skin under varying lighting conditions. The assumption made here is that, apart from the face, no other skin like objects are present in the image. In Rowley et al. [10], use neural networks to detect frontal faces. They do this by using a sliding window over the image and applying a neural network to decide if the image contains a face or not.

In Osuna et al. [7], use a method that utilises a large set of data to train an SVM algorithm which enables them to identify faces. SVM can detect faces by scanning the image for face-like patterns at different scales. This can be used as a core classification algorithm to determine face and non-face objects. In Ping et al. [9], use the template

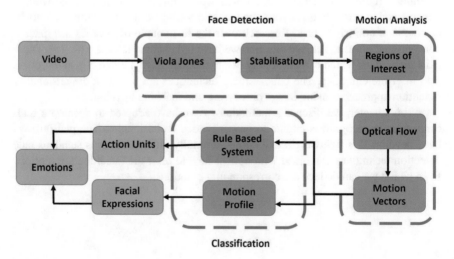

Fig. 2.1 Block diagram showing the key components of the computational framework for facial emotion recognition

matching on multi resolution images to develop a face template which can be utilised to enhance face detection. They then add colour segmentation to identify the face and non-face classes using different types of colour spaces. Similarly, Viola and Jones [13] identify the face by applying a combination of different types of algorithms [1]. These include integral images, Haar like features, Cascade filter and Ada-boost.

In addition to face detection, many applications require to identify and track facial landmarks—the prominent areas of the face such as the eyebrows, eyes, nose and mouth. Such applications include face recognition and video processing as discussed in [8]. In Milborrow and Nicolls [6] use active shape model (ASM). ASM is a statistical model of shapes that predicts the appearance of the objects in new images. Moreover, it identifies facial features by using the training datasets. This process starts by manually identifying the facial feature landmarks, which then uses PCA to determine the variations in the training dataset. This is referred to as a "shaped model".

In Cootes et al. [3], enhances the ASM and presents the AAM. AAM represents shapes which are statistically based on the grey level appearance of the object. This can identify objects, including faces, in various lighting conditions. Both AAM and ASM use PCA to identify the variations in the training datasets. However, the main difference between them is ASM seeks to identify the shape of facial features using points by means of constructing statistical models of them. On the other hand, AAM seeks to match position model and represents the facial feature texture in the image [4]. In Viola and Jones [13] use Haar like features to identify the location of facial features. For this, training datasets are used to identify the location of important areas of the face such as the eyes, mouth and nose [1].

An important point to highlight here is that all of the approaches for facial feature detection, that we are aware of, mostly use the whole face region. However, the essential idea is to identify facial feature regions which are located in different parts of the face. From the extensive experimentations we have performed, we deduce that scanning the entire face usually leads to slow performance and less accuracy. Here we describe a strategy we have developed to process the face and the individual prominent features efficiently.

The first phase in our computational framework is face detection. We use the Viola Jones algorithm for initial face detection which is then corrected using a stabilisation mechanism. The Viola Jones algorithm uses cascade classifiers which are strong classifiers composed of linear weighted weak classifiers. Detecting the face is done by combining these weak classifiers which produce a strong classifier. The combined computation of these weak classifiers returns slightly different face locations each time the algorithm is applied. As a result, it will affect the motion analysis algorithm as it will be considered as a false movement. To solve this problem, we compute a facial normalisation parameter (*FNP*). *FNP* is used to stabilise the detection window in the Viola Jones algorithm and can be computed using Eq. (2.1) such that,

$$FNP = \|loc_i - loc_{(i+1)}\|, \tag{2.1}$$

where, $FNP <= \varepsilon$.

Fig. 2.2 The 4 main ROI
containing the most
prominent facial features.
A—right eye, B—left eye,
C—nose and D—mouth

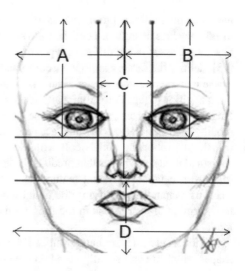

Here, loc_i represents the face location retrieved from the frame i and $loc_{(i+1)}$ represents the face location retrieved from the next frame $i + 1$. FNP can be computed by subtracting the location of both faces and taking the absolute value. To detect the stable face location, we compare the FNP with the threshold value ϵ which can be computed from the size of the face detected compared to the video frame and the frame rate of the video being analysed.

As discussed earlier, most facial feature detection algorithms process the face as a whole leading to poor results. Even applying supervised machine learning algorithms to the whole face, accuracy and performance can be low. This is especially the case when real-time applications are concerned. To address this problem, the strategy we use is to minimise the search area by defining specific ROI. ROI, in this case, are image sections that contain one or more of specific facial features. Our approach comprises of four main steps.

Step 1: Face detection. We use the Viola and Jones [13] algorithm for face detection. After detecting the face we crop it. The output of this step is a left corner point (X_s, Y_s) with a specific width and height which define the bounding box that contains the face in question.

Step 2: Identification of the face centroid. Having successfully located the face from the image, we then divide the face into 4 main parts. This is done by computing the centre point of the face. Usually, the location of the nose appears to be the centre point of the face.

Step 3: Face segmentation. Using the face data available in the CK+ database, we found that the ROI for each of the most prominent facial features are located in one of the four sections of the face as shown in Fig. 2.2. Note, the sections A, B, C and D correspond to the areas containing the facial features, left eye, right eye, nose and mouth respectively.

Table 2.1 Boundary definition of the 4 main ROI in face detection and processing

ROI	Facial feature	Start	End
A	Right eye	(X_s, Y_s)	$((X_s + w)/2, (Y_s + h)/2)$
B	Left eye	$((X_s + w)/2, Y_s)$	$((X_s + w), (Y_s + h)/2)$
C	Nose	$(3/5(X_s + w), Y_s)$	$5/8((X_s + w), (Y_s + h))$
D	Mouth	$(X_s, 3/5(Y_s + h))$	$((X_s + w), (Y_s + h))$

Face segmentation starts by identifying the face location from Step 1, where (X_s, Y_s) is the left corner point of the face location in the image. Additionally, w is the width and h is the height of the image. Thus, X_s, Y_s, w and h are obtained using the Viola Jones algorithm from Step 1. Table 2.1 shows the coordinates of the four main ROI and their boundaries. Here A and B represent the location of the upper quarter of the face, containing the left and right eye. Similarly, C and D represent the location of the nose and mouth respectively.

Step 4: Applying the cascade filter for facial feature detection. For each of the areas A, B, C and D we use the corresponding Viola Jones cascade filter, which can be trained to identify the corresponding facial features. To detect the eyes, we trained three cascade filters, one for detecting both eyes and the other two to detect the left eye and right eye separately. We found that splitting the region of the face containing the eyes into two parts achieved better results than looking for both eyes simultaneously. During real time tracking, it is easy to lose track of both eyes if you have lost track of one of them. Thus, by dividing the region of the faces containing the eyes into two parts, we increase the accuracy of eye tracking. Note, the region labelled as C contains the nose area. By applying a cascade filter to this area we can track the nose accurately. In addition, we found that the nose area could be tracked accurately even when we eventually lose track of the location of the nose.

2.2.2 Motion Analysis

The second phase in our computational framework is the motion analysis routines. This is composed of the identification of the ROI, application of dense optical flow and the computation of motion vectors.

Figure 2.3 shows how we would typically allocate the ROI across the face. We divide the face into three main regions: the upper, middle and lower areas. The upper area includes eyes and eyebrows, the middle area includes the cheeks and the nose, and the lower area contains the mouth and the chin. To capture different movements within each facial feature, we apply a set of ROI over each facial region, where a given ROI represents an approximate location for different parts of facial features.

Fig. 2.3 Typical set of facial regions of interest considered within the computational framework. These regions are broadly based on the underlying facial muscle structure

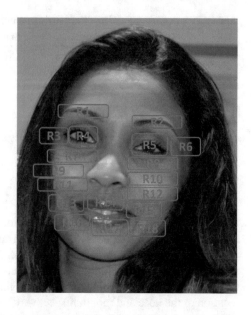

The choice of ROI is decided based on the location of facial muscles. To do this, we studied the facial muscle anatomy and how the muscles are connected to each other. We also studied the literature on how EMG triggers various facial muscle movement which gave us insight into the facial muscular structure and how we should allocate our ROI.

To detect the motion on the face, we apply a dense optical flow algorithm by Farnebäck [5]. An optical flow algorithm is usually applied to image sequences that have a small time gap between them to identify pixel displacements or image motion. Optical flow works under two assumptions: the pixel intensities of an object do not change between consecutive frames and the neighbouring pixels have a similar motion.

Based on this assumption, we need to track pixel values through the image sequence. We need to compute the $I(x, y, t)$ where I is the image intensity value at the location (x, y) at time t which is used to distinguish between the reference frame and current frame. To compute the optical flow in the image sequence, we need to compute the motion between the current frame and the next frame in time dt.

Farnebäck's approach to optical flow computation assumes that all pixels are available in an image to identify the relevant motion. The approach adopted is to approximate the neighbourhood using quadratic polynomial expansions. The motion of a given frame can be expressed as a local "signal" model and within a local coordinate system such that,

$$f(x) = x^T A x + b^T x + c, \qquad (2.2)$$

where, A is a symmetric matrix, b is a vector and c is a scalar. The coefficients are estimated by a weighted least square fit to single pixel values around a given neighbourhood. Based on the hypothesis, if a polynomial undergoes an ideal translation taking into consideration the exact quadratic polynomial form such that,

$$f_1(x) = x^T A_1 x + b_1^T x + c_1, \tag{2.3}$$

a new signal f_2 can be constructed by a global displacement d, such that,

$$f_2(x) = x^T A_2 x + b_2^T x + c_2. \tag{2.4}$$

To estimate the displacement value d, the neighbourhood pixels of the point (x, y) and $(x + dx, y + dx)$ can be considered.

The optical flow algorithm produces orientation and magnitude of each pixel in each ROI. We will need to determine the connection between these motion vectors across various ROI. We describe below how we have achieved it.

To compute the connection between the motion vectors and the various ROI, we use a voting system and a motion converter. The voting system is responsible for averaging the movement over the ROI and eliminating any anomaly caused by different factors like errors in computing the flow and any noise. The voting system consists of a web of connected ROIs.

2.2.3 Classification

For classification, we use two approaches: a rule-based system and the motion profile. The rule-based system uses a set of hard coded rules which are derived from the FACS system. These rules are essentially based on connecting the EMFACS units with the ROI. We undertook several trial and error experiments to devise the connections and to fine tune the rule based system.

Similarly, the motion profile is based on the motion vectors that come from the previous step which are also grouped based on the EMFACS representations and the limits of the underlying muscle motions.

As a result, the computational platform we have developed, though computationally lightweight, can infer facial expressions as well as the relevant FACS coding in an automated fashion.

2.3 Implementation

To implement our approach, we used the OpenCV platform. OpenCV [2] is a free source software with an extensive library consisting of more than 2,500 implemented and optimised algorithms, encompassing classic and state of the art computer vision

(a)   **(b)**

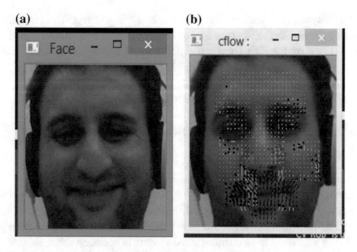

Fig. 2.4 Graphical user interface created using OpenCV and C++. **a** Face detection using Viola Jones. **b** Optical flow computation over the face

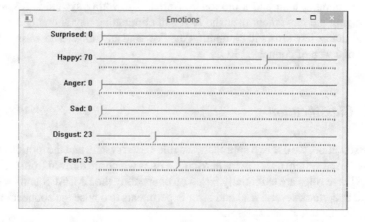

Fig. 2.5 Typical output of the emotion profile visualised using our GUI

and machine-learning algorithms. It is used widely by many research groups, companies and government agencies.

We created a GUI using OpenCV coupled with C++ for real time analysis of videos for generating emotional profiles.

Figure 2.4 shows part of our GUI. Figure 2.4a is the face extracted using the Viola Jones algorithm and Fig. 2.4b shows the various ROI and the optical flow map applied over the resized face.

Figure 2.5, shows a typical output emotion profile from a few video frames. In this example shown, facial expressions are classified as follows: 70% happiness, 23% disgust and 33% fear. The combined form with mixed emotions is usually what we get when we analyse facial emotions. It is important to note that due to the complexity of emotions, it can be extremely difficult to output a pure single basic emotion and so the values we obtain are always mixed. However, for practical purposes, we always take the highest emotional status, if it reaches above a certain threshold value.

2.4 Results

We tested our approach on the CK+ dataset. Figure 2.6 shows the rate of AU detection using our proposed methodology. As can be observed, by excluding certain AUs such as the AU10, AU15 and AU16, the average detection rate of 16 AUs using our approach nears to 93%. Figure 2.7 shows how our system fares against a human FACS coder. Again, our proposed method shows 86% agreement with the professional human FACS coder.

We ran further tests using our system to infer its capacity to produce emotions. Table 2.2 shows our results. As can be observed, our system is fairly accurate when it comes to classifying emotion from video clips.

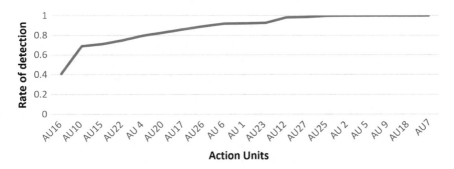

Fig. 2.6 Action unit detection rate using our computational engine

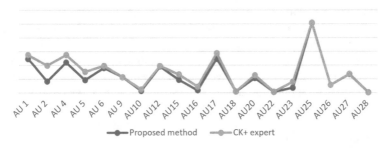

Fig. 2.7 Action unit detection using our computational engine against that performed by a professional FACS coder on the CK+ dataset

Table 2.2 Examples of some basic facial emotions and the corresponding EMFACS coding

Facial emotion	Accuracy (%)
Happiness	95
Surprise	95.2
Sadness	82
Fear	90.5
Anger	90
Disgust	71

2.5   Conclusion

This chapter presents how an automated lightweight facial expression recognition system can be developed. There has been a fair amount of research recently into this topic to try solving this problem by either detecting landmarks or through shape identification using the facial features such as eyebrows, eyes, nose and mouth. The approach we have presented is to identify the regions of facial features and try to analyse the changes (or motion) in these regions using the dense optical flow algorithm.

Here, we detect faces using the Viola Jones algorithm and normalise the face detection method using the FNP in order to increase stability. Furthermore, we identify the location of facial features using ROI. ROI location is created based on studying the facial muscle structure. Then, using optical flow we can detect facial changes during emotional expressions. For classification, we use a rule-based system and motion profile, gaining a detection rate of 86% for 19 action units and 88% correct classification rate for the six basic emotions.

The computational platform presented here comes with its own drawbacks. For example, face tracking is prone to errors and failure when extreme head movement such as large face rotation and head movements are present. Due to the very nature of the optical flow algorithm being implemented, the system has poor performance in extreme lighting conditions such as in very bright environments. In general, these issues are very common in computer based automated facial detection and tracking systems and is an active area of research.

Nonetheless, the proposed automated facial emotion analysis system we have discussed in this Chapter has the advantage of it being fast and less cumbersome. Prominent areas of application using this system can be face analysis in general, for example for deceit analysis, e.g. [16–18]. And, of course, the system we have discussed can be utilised for various aspects of smile analysis. This is discussed further in the next Chapters of this book.

References

1. Al-dahoud, A., Ugail, H.: A method for location based search for enhancing facial feature detection. In: Angelov P., Gegov A., Jayne C., Shen Q. (eds.) Advances in Intelligent Systems and Computing, pp. 421–432. Springer, Cham (2016)
2. About OpenCV, 20/11 (2015). http://opencv.org/about.html
3. Cootes, T.F., Edwards, G.J., Taylor, C.J.: Active appearance models. IEEE Trans. Pattern Anal. Mach. Intell. **23**(6), 681–685 (2001)
4. Cootes, T.F., Edwards, G.J., Taylor, C.J.: Comparing active shape models with active appearance models. BMVC **1999**, 173–182 (1999)
5. Farnebäck, G.: Two-frame motion estimation based on polynomial expansion. In: Bigun, J., Gustavsson, T. (eds.) Image Analysis, SCIA 2003. Lecture Notes in Computer Science, vol. 2749, pp 363–370. Springer, Heidelberg (2003)
6. Milborrow, S., Nicolls, F.: Locating facial features with an extended active shape model. In: Computer Vision, ECCV 2008, pp. 504–513. Springer, Heidelberg (2008)
7. Osuna, E., Freund, R., Girosi, F.: Training support vector machines: an application to face detection. In: Proceedings of IEEE computer society conference on Computer Vision and Pattern Recognition, pp. 130–136. IEEE (1997)
8. Pai, Y-T., Ruan, S-J., Shie, M-C., Liu Y-C.: A simple and accurate color face detection algorithm in complex background. In: 2006 IEEE International Conference on Multimedia and Expo, pp. 1545–1548. IEEE (2006)
9. Ping, S.T.Y., Weng, C.H., Lau, B.: Face detection through template matching and color segmentation. Nevim. Nevim **89** (2003)
10. Rowley, H.A., Baluja, S., Kanade, T.: Neural network-based face detection. IEEE Trans. Pattern Anal. Mach. Intell. **20**(1), 23–38 (1998)
11. Singh, S.K., Chauhan, D., Vatsa, M., Singh, R.: A robust skin color based face detection algorithm. Tamkang J. Sci. Eng. **6**(4), 227–234 (2003)
12. Valstar, M.F., Mehu, M., Jiang, B., Pantic, M., Scherer, K.: Meta-analysis of the first facial expression recognition challenge. IEEE Trans. Syst. Man Cybern. Part B Cybern. **42**(4), 966–979 (2012)
13. Viola, P., Jones, M.J.: Robust real-time face detection. Int. J. Comput. Vis. **57**(2), 137–154 (2004)
14. Yap, M.H., Ugail, H., Zwiggelaar, R., Rajoub, B., Doherty, V., Appleyard, S., Huddy. G.: A short review of methods for face detection and multifractal analysis. In: Cyberworlds 2009, Bradford, UK (2009)
15. Yap, M.H., Ugail, H., Zwiggelaar, R., Rajoub, B.: Facial image processing for facial analysis. In: IEEE International Carnahan Conference on Security Technology (ICCST 2010), California, USA, San Jose (2010)
16. Yap, M.H., Ugail, H., Zwiggelaar, R.: Intensity score for facial actions detection in near-frontal-view face sequences. Comput. Commun. Eng. **6**, 819–824 (2013)
17. Yap, M.H., Ugail, H., Zwiggelaar, R.: Facial analysis for real-time application: a review in visual cues detection techniques. J. Commun. Comput. **9**, 1231–1241 (2013)
18. Yap, M.H., Ugail, H., Zwiggelaar, R.: Facial behavioural analysis: a case study in deception detection. Br. J. Appl. Sci. Technol. **4**(10), 1485–1496 (2014)

Chapter 3
Distinguishing Between Genuine and Posed Smiles

Abstract This chapter presents an application of computational smile analysis framework discussed earlier. Here we discuss how one could utilise a computational algorithm to distinguish between genuine and posed smiles. We utilise aspects of the computational framework discussed in Chap. 2 to process and analyse the smile expression looking for clues to determine the genuineness of it. Equally, we discuss how the exact distribution of a smile across the face, especially the distinction in the weight distribution between a genuine and a posed smile can be achieved.

Keywords Smile analysis · Genuine smiles · Posed smiles · Smile weight distribution

3.1 Introduction

It is widely known that a smile not only can communicate happiness, but it can also communicate other emotions including negative emotions such as fear and disgust [3, 4, 8]. It is understood that the genuine smiles reflect the feeling of enjoyment and elicit pleasure connecting it with social rewards [6, 11]. It also triggers a number of positive emotions as well as cooperative behaviour [7]. On the other hand, posed smiles are considered to be less rewarding which are often used by people to mask their negative feelings [5].

Being able to accurately distinguish between Duchenne and non-Duchenne smiles is often an important task. However, this task, though on the face of it appear to be simple, it is still very challenging [3]. In experiments involving identification of personality, loveliness and humour, it appears that people show the so called Duchenne marker—one often related to the genuine smile. However, it has been shown that AU6, as well as, subtle dynamic properties, the timing involved in the smile and the amount of eye constriction are all critical in identifying a genuine smile.

There has been a fair amount of research done on the topic of distinguishing between genuine and posed smiles. For example, Dimberg [2] used EMG inferred reactions to classify Duchenne and non-Duchenne smiles, identifying the muscles

H. Ugail and A. A. A. Aldahoud, *Computational Techniques for Human Smile Analysis*, SpringerBriefs in Computer Science, https://doi.org/10.1007/978-3-030-15381-6_3

related to these smiles in the context of other emotions such as disgust, fear, anger, sadness and surprise. Similarly, Mai et al. [9] investigated Chinese participants to judge for Duchenne and non-Duchenne smiles by focusing on the mouth and the eyes.

On the computational front, Wu et al. [12] have proposed a computational system to identify the genuineness of a smile based on the detection of AUs 6 and 12, based on utilising PCA combined with Gabor filters and SVM. They report over 85% accuracy in smile classification. Similarly, Nakano et al. [10] utilised PCA along with neural networks to achieve smile classification rates of up to 90%.

Here we focus on to explain how a smile is distributed across the face and how we can distinguish between genuine and posed smiles through the weighted distribution of the smile across various facial features. Next, we discuss a computational framework one can implement to do that. In doing so, we show that the genuine smile is in the eyes. We also show the proportional distribution of the weight of a smile around the mouth and eyes. A computational tool of this nature is useful for devising efficient human computer interaction systems. It also helps one to develop soft biometrics to complement existing biometric authentication systems. We believe such a tool will be equally useful to social and clinical scientists who are keen to have a deeper understanding of the behavioural and personality traits by studying the face.

3.2 A Computational Framework for Smile Weight Distribution

There are 18 distinct types of smiles and all of which must bear the three distinct phases mentioned above, though the effects pronounced on different parts of the face do differ in each case. For example, as discussed earlier, in a genuine smile the corners of the mouth are raised. And more significantly, the narrowing of the eye aperture happens. This results in the appearance of crows feet around the outer sides of the eyes.

As far as the dynamics of a smile is concerned, there are three distinct phases in any type of smile. They are, the onset (which starts from the neutral facial expression to the peak of zygomatic major muscle contraction), the apex (which refers to the time it takes for the smile to stay in the expressive state) and the offset (where the facial expression reaches from expressive state back to the neutral). The discussion here predominantly focusses on the dynamics of the apex part of the smile to categorise whether a smile is genuine or not. Our framework is based upon computing the dynamic geometry changes across the face, through the employment of optical flow analysis.

Figure 3.1 shows a block diagram of the computational framework that one can use to analyse a smile to see whether it is genuine or not. This framework contains three main parts. They are detection, face analysis and finally the output of the

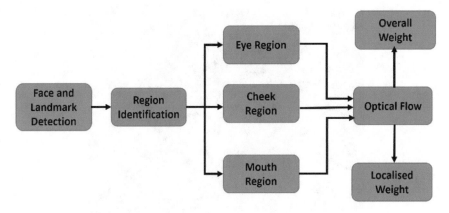

Fig. 3.1 Description of the computational framework for the analysis of the smile dynamics

results. The detection phase includes face detection, face resizing, landmark detection and identifying the ROI. To do this, we use the methodology discussed in Chap. 2, where we use the Viola Jones algorithm along with identification and processing of individual facial features.

Once the face is detected and processed accordingly, it can then be resized to a suitable working size. This is to ensure uniformity is maintained across all the ROI and that landmark detection is undertaken appropriately for each frame of video consisting a face that we must analyse.

There are various mechanisms one can utilise to detect individual landmarks on the face. In this case, we use the CHEHRA model [1], which is a machine learning algorithm used to detect the facial landmarks as shown in Figs. 3.2 and 3.3. The CHEHRA algorithm has been trained to detect facial landmarks using "faces in-the-wild datasets" under varying illumination, facial expressions and head pose. Several experiments using this model demonstrate that it is efficient and it can be utilised for near real time applications.

Upon detection of the landmarks, we define the relevant ROI such as the area around the eyes, cheeks and the mouth. To do this, we first identify all the landmarks in a neutral facial expression which helps us to identify the initial location of the mouth, cheeks and eyes. This process also enables us to normalise the ROI identification process. Details of these ROI are shown in Fig. 2.3 in Chap. 2.

Table 3.1 shows the relevant facial features (i.e. the mouth, eyes and the cheeks) and the associated ROI as well as the corresponding landmarks. The computation of a specific ROI is done through two steps. First, we locate the relevant reference landmarks which are defined as two 2-dimensional vectors (x, y) where variation across each axis is considered separately. Second, using the formulation (λ_x, λ_y) we can denote a shift distance for each axis from the reference points. Thus, using the shift distances (λ_x, λ_y) along with the origin of the central landmark point, we can compute the ROI using Eq. (3.1) such that,

Fig. 3.2   An example of an input face for landmark detection using the CHEHRA model

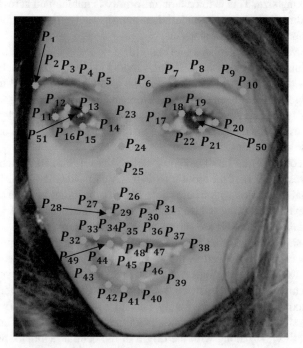

Fig. 3.3   Facial landmarks detected using the CHEHRA model

Table 3.1 Description of the regions of interest, the relevant facial features and the corresponding facial landmarks

Facial feature	ROI	Reference landmarks
Eyes	R1	$P_{1,x}, P_{2,y}$
	R2	$P_{6,x}, P_{2,y}$
	R3	$P_{12,x}, P_{1,y}$
	R4	$P_{11,x}, P_{12,y}$
	R5	$P_{17,x}, P_{18,y}$
	R6	$P_{20,x}, P_{19,y}$
Cheeks	R7	$P_{11,x}, P_{16,y}$
	R8	$P_{17,x}, P_{22,y}$
	R9	$P_{32,x}, P_{32,y}$
	R10	$P_{38,x}, P_{38,y}$
	R11	$P_{32,x}, P_{32,y}$
	R12	$P_{38,x}, P_{38,y}$
Mouth	R13–R18	$P_{32(X,Y)}, P_{33(X,Y)}, P_{35(X,Y)},$ $P_{37(X,Y)}, P_{38(X,Y)}, P_{39(X,Y)},$ $P_{41(X,Y)}, P_{42(X,Y)}$

$$ROI = \begin{cases} RO_x = P_x \pm \lambda_x \\ RO_y = P_y \pm \lambda_y \end{cases}, \tag{3.1}$$

where P_x is the reference point along the x-axis and P_y is the reference point along the y-axis.

It should be noted that there are ROI where no obvious boundaries can be defined such as the R9, R10, R11, R12 and that below the eyes, R7 and R8. In such cases, where landmarks cannot be directly identified, we use the Euclidean wise nearest reference landmark. For example, to locate the left eyebrows, using landmark P_1 we allocate a window of appropriate size (110 by 35 pixels in this case) which covers the area surrounding the eyebrows. Similarly, to locate the right cheek, we use the mouth right corner point P_{38} as the reference landmark to infer R12, R10 and R8.

Using the above formulation and by means of using the facial landmarks as reference points, we thus allocate the ROI. Moreover, we identify 18 ROI through which the motion around mouth, cheeks and eyes can be monitored. In particular, we identify 4 different ROI around the eye, i.e. eyebrow (R1 and R2), eye corners (R4, R5) and (R3, R6) and the area beneath the eye (R7, R8). The reason we allocate such greater number of ROI to the eye area is that we wanted to study the regions around the eye area in greater detail. This is because previous work has shown that there are greater distinctions in the level of activity around eye area between genuine and non-genuine smiles.

Once the relevant ROI are allocated its motion over time (through the dynamics of a smile) can be tracked and analysed. To do this, we apply the optical flow algorithm discussed in Chap. 2. Once the optical flow values are computed, they need to be normalised in order to overcome some of the challenging factors such as the face location relative to the camera, the changing size of the face as it moves from one frame to the other. To obtain a uniform normalisation, we use one of the most stable areas of the face namely, the triangular area formed with the tip of the nose, P_{29} and the two eye corners, P_{11} and P_{20}.

Once we compute the optical flows we can then compute the relevant smile weight distribution. The weight distribution of a given smile is computed in two ways, i.e. finding the localised weight distribution and the overall weight distribution. The overall weight distribution represents the relation between all the main facial features in genuine and posed smiles. Similarly, the localised weight distribution represents the relation between different parts of the facial features in genuine and posed smiles. Determining the overall weight is based on computing the flow for each region of the face and summing these flows in an appropriate way. Similarly, the localised flow is achieved by computing the flow in specific ROI and then summing up all flows in each possible direction which represents the total displacement in specific ROI.

3.3   Results

In this section, we discuss the results of some of the experiments to show how the framework proposed earlier can be utilised to distinguish between genuine and posed smiles. Further, we show that it can also be utilised to know the exact distribution of the weight of a smile over an individual feature of the face. Here, we make use of the CK+ dataset with 82 subjects expressing posed smiles and the MUG dataset with 52 subjects expressing induced genuine smiles.

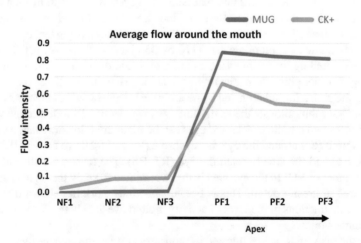

Fig. 3.4 Average flow around the mouth for the subjects in the MUG and the CK+ datasets

First, we note that the flow analysis between the neutral to peak smile expressions is interesting to study. To do this, we measure the flow value for the first three frames and the last three frames of the smile for both datasets. The first three frames are denoted by NF_1, NF_2 and NF_3 which represent the neutral expressions or the start of the smile. The last three frames are denoted by PF_1, PF_2 and PF_3 which represent frames in the peak of the smile. Using the optical flow algorithm described previously, we measure the flow which represents the displacement value of pixels in the related ROI for each of the facial features. In order to check if these values have a significant meaning, we can compute the average of each frame in neutral and peak frames for all the subjects in the datasets. Figures 3.4, 3.5 and 3.6 show the average of the neutral

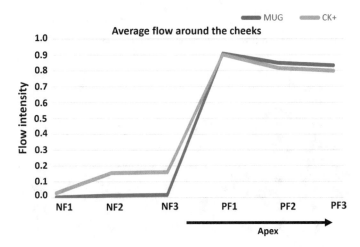

Fig. 3.5 Average flow around the cheeks for the subjects in the MUG and the CK+ datasets

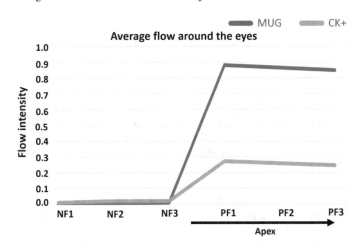

Fig. 3.6 Average flow around the eyes for the subjects in the MUG and the CK+ datasets

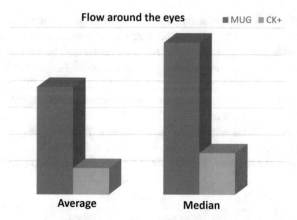

Fig. 3.7 Average and median flow around the eyes for the subjects in the MUG and the CK+ datasets

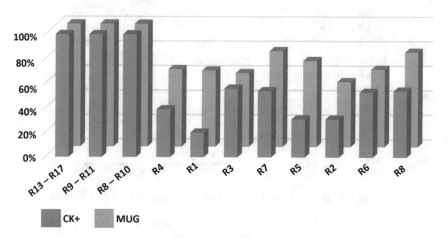

Fig. 3.8 Smile weight distribution for individual ROI for the subjects in the MUG and the CK+ datasets

and peak frames for mouth and cheeks. The results in the figure do indicate that the MUG dataset has a higher average compared to the CK+ dataset. Furthermore, in Fig. 3.7, we show both the average and median flow around the eyes for the subjects in the MUG and the CK+ datasets. Again, it shows that the flow around the eyes is higher for the subject in the MUG dataset.

Furthermore, in Fig. 3.8, we show the smile weight distribution for individual ROI for the subjects in the MUG and the CK+ datasets. In Fig. 3.9, we show the smile weight distribution around the mouth, cheeks and the eyes for the subjects in the MUG and the CK+ datasets. Finally, in Fig. 3.10, we show heat maps on faces to help visualise the smile distribution between a genuine and a posed smile.

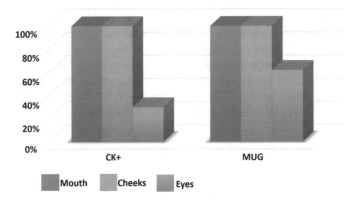

Fig. 3.9 Smile weight distribution around the mouth, cheeks and the eyes for the subjects in the MUG and the CK+ datasets

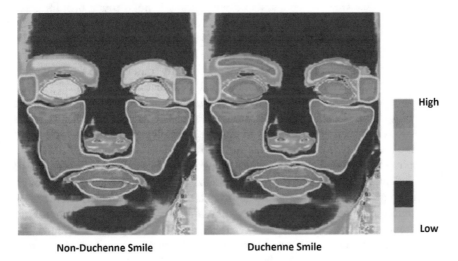

Fig. 3.10 Heat maps on faces to help visualise the smile distribution between a genuine and a posed smile

3.4 Discussions

As can be seen from the above results, we found there is a significant difference in the flow values and movement occurrences between Duchenne and posed smiles. In the first experiment, where we looked at the first three frames, NF_1, NF_2 and NF_3, and the last peak frames, PF_1, PF_2 and PF_3 of the smile, the results show that both the average and the median for the flow around the facial features are significantly different. Furthermore, the results also indicate that there is a significant difference between the flow around the eyes for genuine and posed smiles. In fact, in some cases, we observed a 4-fold increase in the average flow values around the eyes in the case for the genuine smile. Additionally, we computed the flow distribution for the subjects expressing genuine and posed smiles for each of the related ROI in the areas around the eyes. The results show that subjects expressing genuine smiles have a higher average flow distribution and higher median values in the case of a genuine smile. In general, from the experiments we have undertaken, we can assume there is almost 10% higher rate of activity around the eye in the case of a genuine smile. This measure thus can be broadly applied for distinguishing between a genuine and a posed smile.

3.5   Conclusions

A smile is one of the universal facial expressions observable across humans, yet distinguishing between a genuine and a polite smile is not a trivial task for us to perform. In fact, it is true to state that, regardless of the type of smile we are observing, our gaze often would get fixated around the mouth area of someone while she or he is smiling. However, research suggests that the hints for a genuine smile do not lie around the mouth area of a person but are prominently found around the regions of the eyes. More specifically, it is postulated that the orbicularis oculi muscle, which rings the eye, produces significant tension during a genuine smile when compared to a polite smile. This activation of orbicularis muscle in genuinely happy facial expressions has been noted by both in observed human experiments as well as by the more invasive type of experiments such those carried out using facial EMG.

Inspired by such studies, in this research, we set out to study the human smiles in a detailed manner through the use of computer based non-invasive analysis. Through the employment of state of the art image analysis techniques, we undertook the automated analysis of the motion of various parts of the face. As a result, we demonstrate the presence of significant facial activity around the eyes during a genuine smile. We argue that the employment of a computer based image analysis tools can significantly enhance automated non-invasive facial analysis.

This work can be taken further forward in various directions. One direction we feel this work should be taken forward is the detailed analysis of the smile dynamics through extensive experimentation using the framework we have described. For example, in this case, we just studied the flow intensity during the apex phase of the smile. It would be interesting to look at the onset as well as the offset phases of the smile too. Further, it will be interesting to use more sophisticated analysis techniques, beyond simple averaging and median computations, for a deeper understanding of the dynamics of smiles and their detailed weight distributions.

References

1. Asthana, A., Zafeiriou, S., Cheng, S., Pantic, M.: Incremental face alignment in the wild. In: CVPR 2014 (2014). https://ibug.doc.ic.ac.uk/resources/chehra-tracker-cvpr-2014/
2. Dimberg, U.: Facial electromyographic reactions and autonomic activity to auditory stimuli. Biol. Psychol. **31**(2), 137–147 (1990)
3. Ekman, P.: Universal and cultural differences in facial expression of emotion. In: Nebraska Symposium on Motivation, 1971, pp. 207–283. Nebraska University Press, Lincoln (1972)
4. Ekman, P.: Cross-cultural studies of emotion. In Darwin and facial expression: a century of research in review, pp. 169–222. Academic Press, New York (1973)
5. Ekman, P., Davidson, R.J., Friesen, W.V.: The Duchenne smile: emotional expression and brain physiology II. J. Pers. Soc. Psychol. **58**, 342353 (1990)
6. Frank, M.G., Ekman, P., Friesen, W.V.: Behavioral markers and recognizability of the smile of enjoyment. J. Pers. Soc. Psychol. **64**(1), 83–93 (1993)
7. Krumhuber, E., Manstead, A.S.R., Cosker, D., Marshall, D., Rosin, P.L., Kappas, A.: Facial dynamics as indicators of trustworthiness and cooperative behavior. Emotion **7**, 730735 (2007)
8. LaFrance, M.: Lip service: Smiles in life, death, trust, lies, work, memory, sex, and politics, Norton & Company (2011)
9. Mai, X., Ge, Y., Tao, L., Tang, H., Liu, C., Luo, Y.-J.: Eyes are windows to the Chinese soul: evidence from the detection of real and fake smiles. PloS One **6**(5), e19903 (2011)
10. Nakano, M., Mitsukura, Y., Fukumi, M., Akamatsu, N.: True smile recognition system using neural networks. In: Proceedings of the 9th International Conference in Neural Information Processing, ICONIP'02, pp. 650–654. IEEE (2002)
11. Shore, D.M., Heerey, E.A.: The value of genuine and polite smiles. Emotion **11**, 169174 (2011)
12. Wu, P., Wang, W., Liu, H.: Methods of recognizing true and fake smiles by using AU6 and AU12 in a holistic way. In: Proceedings of 2013 Chinese Intelligent Automation Conference, pp. 603–613. Springer, Heidelberg (2013)

Chapter 4
Gender and Smile Dynamics

Abstract This chapter is concerned with the discussion of a computational frame-work to aid with gender classification in an automated fashion using the dynamics of a smile. The computational smile dynamics framework we discuss here uses the spatio-temporal changes on the face during a smile. Specifically, it uses a set of spa-tial and temporal features on the overall face. These include the changes in the area of the mouth, the geometric flow around facial features and a set of intrinsic features over the face. These features are explicitly derived from the dynamics of the smile. Based on it, a number of distinct dynamic smile parameters can be extracted which can then be fed to a machine learning algorithm for gender classification.

Keywords Smile dynamics · Gender recognition · Machine learning · k-nearest neighbour

4.1 Introduction

Automated gender classification has numerous applications ranging from human-computer interaction to soft biometrics. Reliable, as well as, computationally efficient gender recognition from a video is a challenging task and is an active area of research [3]. This is due to the algorithmic computational challenges relating to the ethnicity and the age of the person, the varying lighting conditions of a given environment as well as the problems of facial poses and occlusion issues, e.g. [7, 10, 12–14].

The majority of work relating to image assisted computational gender classifica-tion available in the literature uses appearance based approaches—the goal being to find discriminating features on the face between females and males. However, this is not a very trivial task since it is not entirely conclusive that females and males have entirely distinct facial features to compare and contrast with [8].

Computational gender classification methods using images usually contain two major components. They are a feature extraction phase followed by a pattern clas-sification phase. For feature extraction a number of image processing and analysis techniques are available. Some of them include, SIFT, AAM and LBP. Additionally,

© The Author(s), under exclusive licence to Springer Nature Switzerland AG 2019   35
H. Ugail and A. A. Aldahoud, *Computational Techniques for Human Smile Analysis*,
SpringerBriefs in Computer Science, https://doi.org/10.1007/978-3-030-15381-6_4

combinations or modifications of these techniques are also utilised. For classification, both rule based and machine learning techniques are employed.

The discussion of this Chapter focusses on gender classification using the dynamic components of a smile. Similar work on gender and smiles also provide clues as to how this may be feasible [5]. Determining the gender itself from a dynamic face has many applications. For example, there might be cases where one might be presented with grainy or blurry video footage of a subject for identification and on the face of it using standard face recognition it may not be possible to recognise the individual. However, there may be enough dynamic information from certain regions of the face which might be useful in knowing the gender of the person in question which then may lead to the determination of the identity of that individual. Other potential applications of accurate automatic gender recognition may be, automated counting of the number of women or men in a public place or developing human machine interaction systems which can automatically grant access to certain areas based on the gender [6].

Thus, here we discuss a novel technique for gender estimation based on the spatio-temporal components of the smile. To do this, we show that one can utilise a number of smile features based on the changes in the mouth, the geometric flow around facial features of the face and also a set of intrinsic features derived from the dynamic geometry of the face.

4.2   A Computational Framework for Smile Dynamics

Both social and psychological experiments show strongly that there exist traits of gender in the smile. The computational framework discussed here is based on four key components. They are, (1) the dynamic geometric distances which we can refer to as the spacial parameters, (2) the dynamics of the mouth representing the changes in the area of mouth during a smile, (3) the optical flow computed around the key parts of the face, and (4) some features which are intrinsic to the dynamic geometry of the face. The motivation for choosing these dynamic features of the smile expression is that there exists ample literature to show that such dynamic features of the smile are stable, e.g. [2, 4].

Figure 4.1 presents the proposed computational framework to analyse the dynamic components of the smile. We first detect and trace the face from a given video sequence. To do this, we have used the Viola Jones algorithm along with local detection and processing of key facial features from the face as discussed in Chap. 2, and references [1, 9]. Again, we utilise the CHEHRA model for automatic facial landmark detection as discussed in Chap. 3. From the CHEHRA algorithm, we detect 49 landmarks on the face, marked as $P_1 ... P_{49}$, as shown in Fig. 3.3.

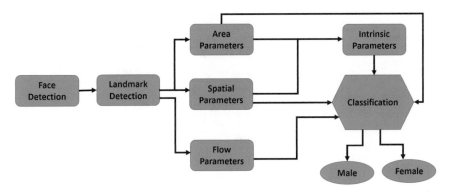

Fig. 4.1 Block diagram showing the key components of the computational framework for automatic analysis of the dynamics of a smile

4.2.1 Dynamic Spatial Parameters

Using the CHEHRA model, we obtain 49 facial landmarks. We then identify 6 Euclidean distances across the face from which the spatial parameters can be derived. This is then used to compute the dynamic spatial parameters. Details of these dynamic spatial parameters are given in Table 4.1. We describe the general form of a given spatial parameter as,

$$\delta d_i = \frac{d_i}{N_i} + \sum_{n=1}^{t} \frac{d_i}{N_i} - \frac{d_{in}}{N_{in}}. \tag{4.1}$$

Here t is the total number of video frames corresponding to each $\frac{1}{10}$th increment of the entire time period T for the smile—running from neutral expression to the peak of the smile. N_i is the distance of the nose. It is computed using the distance between P_{23} and P_{26}. Hence, by dividing each of the spatial parameters by the length of the nose N_i, one can normalise these parameters to the given dynamic facial image. Note that, for a given smile, from neutral to the peak, we divide the time it takes into ten partitions. Therefore, for each of the d_i one can have 10 times d_i parameters. These parameters can then be fed to the machine learning algorithm. Thus, in this case, we have a total of 60 dynamic spatial parameters.

Figure 4.2 shows the variation of δd_i across the 10 time partitions. This is for a typical smile. It can be seen, there is a variation in each parameter as the smile progresses from neutral expression to its peak.

Table 4.1 Description of the dynamic spatial parameters

Spatial distance	Description	Corresponding landmarks
d_1	Corners of the mouth	P_{32} to P_{38}
d_2	Lower and upper lip	P_{45} to P_{48}
d_3	Mouth to nose (left side)	P_{32} to P_{27}
d_4	Mouth to nose (right side)	P_{38} to P_{31}
d_5	Eye to Mouth (left corners)	P_{32} to P_{11}
d_6	Eye to Mouth (right corners)	P_{38} to P_{20}

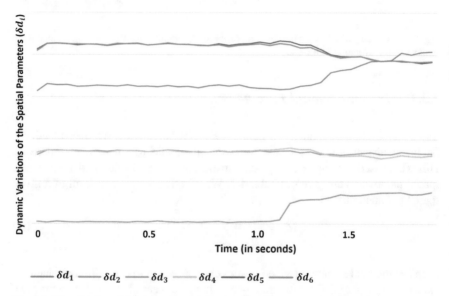

Fig. 4.2 Visualisation of the dynamic spatial parameters δd_i across the 10 partitions of time. This is for a typical smile, starting from the neutral expression to the peak of the smile

4.2.2   Area Parameters on the Mouth

The second set of dynamic parameters are identified on the mouth. To do this, one can compute the changes in the area of 22 triangular regions that occupy the total area of the mouth. This is shown in Fig. 4.3. Like previously, the areas are computed based on the corresponding landmarks which are obtained from the CHEHRA model. The general form of how the dynamics in the mouth area are computed is described as,

$$\Delta_{area}^{i} = \sum_{n=1}^{22} \frac{\Delta_i}{\Delta N_i}, \qquad (4.2)$$

and,

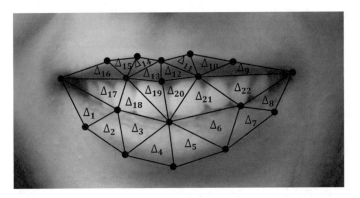

Fig. 4.3 Description of triangular areas of the mouth which is used to formulate the dynamic area parameters on the mouth

$$\delta\Delta_i = \sum_{n=1}^{t} \Delta_{area}^i. \tag{4.3}$$

Here, t is the total number of video frames corresponding to each $\frac{1}{10}th$, corresponding to the incremental changes on the total time T for the smile, from neutral expression to the peak of the smile. The ΔN_i here is the invariant triangle area determined by the landmarks defining the outer the tip of the nose and the outer corners of the eyes. These are marked as P_{11}, P_{20} and P_{26}. Similar to identifying the spatial parameters, we divide the total time of the smile, from neutral to peak, into ten partitions. Thus, we obtain 10 parameters from the $\delta\Delta_i$, through time. In Fig. 4.3 we show the distribution of areas of the triangular regions. For the purpose of illustration, in Fig. 4.4, we show the variations through time of these parameters for a smile.

4.2.3 Geometric Flow Parameters

The computation of geometric flow around the face is the next component in our smile dynamics framework. For this, we compute the flow around the mouth, cheeks and around the eyes. This is done using the optical flow algorithm discussed in Chap. 2. Using this algorithm, one is able to estimate the successive displacement of each of the landmarks during the smile. Table 4.2 shows the landmarks and regions of the face that can be used for computing the geometric flow around the key features of the face. We show the corresponding facial regions and landmarks in Fig. 4.5. Figure 4.6 shows the variations in the dynamic optical flows, δf_i, around the face for a typical smile.

Once we compute the flow, δf_i, for each of the regions, it is important to normalise them upon computation. To do this, one can use the corresponding flow around the

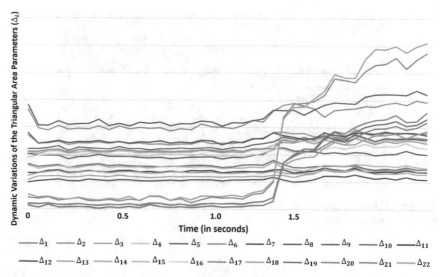

Fig. 4.4 Variation in the dynamic area parameters Δ_i on the mouth, across the 10 intervals of time, for the smile, from neutral expression to the peak of the smile

Table 4.2 Derivation of the optical flow parameters around the face

Optical flow	Description	Landmarks/regions
δf_1	Mouth area	Landmarks P_{32} to P_{49}
δf_2	Left eye area	f_6, f_7, f_8, f_9
δf_3	Right eye area	f_1, f_2, f_3, f_4
δf_4	Left cheek area	f_{10}
δf_5	Right cheek area	f_5

invariant triangle area of the face. This triangle is determined by using the landmarks P_{11}, P_{20} and P_{26}. The geometric flow parameters, δf_i, is computed across the 10 incremental time intervals. This results in a total of 50 dynamic geometric flow parameters which are then fed to machine learning.

4.2.4 Intrinsic Parameters

Here, we compute a family of intrinsic dynamic parameters on the face. This enables one to further enhance the analysis of the dynamics of the smile. Note, that the intrinsic parameters are based on the computation of the variations in the slopes. They are also based on the growth rates of various other features across the face. These features are described as s_1, s_2, s_3 and s_4.

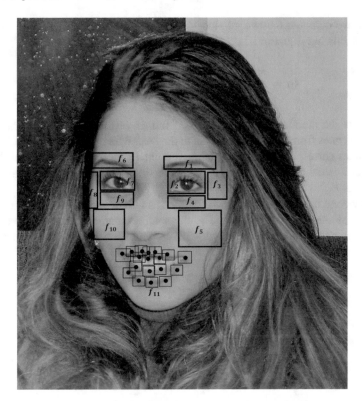

Fig. 4.5 Regions of the face identified for dynamic optical flow computation

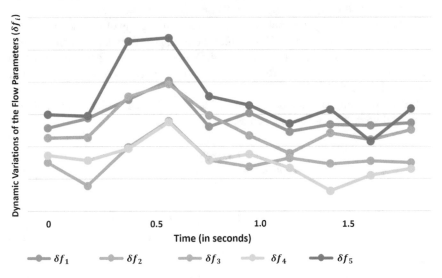

Fig. 4.6 Variations in the dynamic optical flows δf_1 around the face, for a typical smile, from neutral to the peak

First, we compute the overall slope variation around the mouth. To do this, we use the following Equation.

$$s_{1i} = \frac{N \sum_{n=1}^{N} P_{ix} P_{iy} - \sum_{n=1}^{N} P_{ix} \sum_{n=1}^{N} P_{iy}}{\sum_{n=1}^{N} P_{ix}^2 - (\sum_{n=1}^{N} P_{ix})^2}. \tag{4.4}$$

Here N is the number of frames of video frames of the entire clip of the smile. Again this goes from neutral expression to the peak where P_{ix}. P_{iy} are the Cartesian coordinates corresponding to the landmark point P_i. In this case, we obtain a total of 12 parameters considering the variations in slopes around mouth relating to the landmarks P_{32} to P_{43}.

The second family of parameters is defined as s_2. These correspond to the growth rates across the smile. These relate to the spatial parameters as well as the area parameters on the mouth. The growth rates coming from the spatial parameters are defined as,

$$s_{2i(spatial)} = \sum_{n=1}^{N} \frac{\delta d_i^t - \delta d_i^{t+1}}{\delta d_i^t}. \tag{4.5}$$

The area parameters on the mouth are,

$$s_{2i(area)} = \sum_{n=1}^{N} \frac{\Delta_i^t - \Delta_i^{t+1}}{\Delta_i^t}. \tag{4.6}$$

Here N is identified as the total number of frames in the video sequence, and t to $t + 1$ defines two corresponding video frames. The growth rates $s_{2i(area)}$ are then computed. Additionally, for each of the 22 triangular regions of the mouth we also compute the total growth rate. This defines the total growth rate of the entire smile for the mouth. This is done by using Eq. (4.6) along with the 22 triangular mouth area information. This means, for the dynamic intrinsic type s_2, we have a total of $6 + 22 + 1 = 29$ parameters.

The third family of parameters we define is s_3. We can refer to these as the compound growth rates and is given as,

$$s_{3i(spatial)} = \left(\frac{\delta d_i^{neutral}}{\delta d_i^{peak}} \right)^{1/N} - 1, \tag{4.7}$$

and,

$$s_{3i(area)} = \left(\frac{\Delta_i^{neutral}}{\Delta_i^{peak}} \right)^{1/N} - 1. \tag{4.8}$$

Here N, like previously, is the total number of frames in the video sequence of the smile. Note that, the compound growth rate is measured simply using the neutral and

Table 4.3 Parameter description for the computational framework for smile dynamics

Parameter	Description	No. of parameters
δd_i	Spatial—involving 6 geometric distances across the face	60
$\delta \Delta_i$	Mouth area—derived from the total area for the 22 parts of the mouth	10
δf_i	flow around the eyes, mouth and cheeks	50
s_{1i}	Slope around mouth landmarks P_{32} to P_{43}	12
s_{2i}	Growth rates of the mouth area and for the spatial parameters	29
s_{3i}	Compound growth rates of the mouth area and for the spatial parameters	29
s_{4i}	Gradient orientations for the mouth area and for the corners of the mouth	20

peak of the smile expression. Additionally, we also compute the compound growth for the entire mouth. This is done using the total area of the mouth. Hence we obtain a total of 29 parameters of dynamic intrinsic type s_3.

The final family of parameters in this category is s_4. For this, we compute the gradient orientation of the mouth based on the two mouth corner landmarks P_{32} and P_{38}. This provides us with a line m passing δd_1 at the neutral and the peak of the smile. We then use,

$$s_{4i} = \sum_{t=1}^{T} \delta d_1^t - m^t. \tag{4.9}$$

This equation is used to compute the rate of deviation of the mouth corners against the gradient m over the 10 time partitions, Here T is the total time from neutral to the peak of the smile. Similarly, we can compute the gradient orientation of the mouth area. This is based on the combined 22 triangular areas of the mouth.

These parameters can be considered as *smoothing* parameters of the smile. These also form an additional $10 + 10 = 20$ parameters for machine learning.

Table 4.3 provides a description of various parameters associated with our computational framework for smile dynamics.

Table 4.4 Results using the k-NN based machine classification

	CK+	MUG
k-NN distance	Correlation	Cosine distance
k value	3	14
Machine classification	78%	86%

4.3 Classification Using Machine Learning

A number of experiments by social scientists and psychologists do confirm that gender is indeed encoded in the smile. Therefore, it will be interesting to see how such distinction can be derived using the computational framework we discussed above. To do this, one can utilise machine learning algorithms. To demonstrate this, we utilised both SVM and the k-NN algorithms. From the various dynamic smile parameters described earlier, we obtain a total of 210 unique features which can then be utilised to train a classifier.

The k-NN algorithm is a class of machine learning algorithms. It is a non-parametric method that is used for classification and regression. In the case of the k-NN machine classification we have utilised here, we made use of all the 210 features obtained from the dynamics. These are shown in Table 4.3 which are used to train our classifier. To validate the accuracy of our methods we also used a 10-fold cross-validation. In addition to this, we can use a number of distance functions to test the correlations. Examples of such distance functions include Euclidean, Cosine and Minkowsky. In Table 4.4, we report the best results we have obtained.

4.4 Conclusion

In this Chapter, we have discussed how one can consider the analysis of the dynamic face, in particular, the dynamics of the smile, to be a superior alternative for gender classification. The framework uses spatio temporal characteristics of the face to infer gender. Dynamic parameters such as the area of the mouth, intrinsic parameters on the face and optical flow parameters across the face are utilised as part of the computational model presented here. These components can produce a total of 210 unique features. These features can then be fed to a machine learning algorithm for feature learning and classification. It has been observed, during a smile, there is more dynamic activity in the lip area of females in comparison to males. In addition to this, using k-NN based machine learning, one can obtain a gender classification rate of 86%, simply using dynamic components of a smile.

References

1. Al-dahoud, A., Ugail, H.: A method for location based search for enhancing facial feature detection. In: The Proceedings of the International Conference on Advances in Computational Intelligence Systems, AISC, pp. 421–432 (2016)
2. Brody, L.R., Hall, J.A., Stokes, L.R.: Gender and emotion: theory, findings, and content. In: Barrett, L.F., Lewis, M., Haviland-Jones, J.M. (eds.) Handbook of Emotions, 4th edn, pp. 369–392. The Guildford Press (2016)
3. Bukar, A.M., Ugail, H., Connah, D.: Automatic age and gender classification using supervised appearance model. J. Electron. Imaging 25(6), 061605 (2016)
4. Cashdan, E.: Smiles, speech, and body posture: how women and men display sociometric status and power. J. Nonverbal Behav. 22(4), 209–228 (1998)
5. Dantcheva, A., Brémond, F.: Gender estimation based on smile-dynamics. IEEE Trans. Inf. Forensics Secur. 12(3), 719–729 (2017)
6. Han, X., Ugail, H., Palmer, I.: Gender classification based on 3D face geometry features using SVM. In: Cyberworlds 2009, Bradford, UK (2009)
7. Jacko, J.A.: Human Computer Interaction Handbook: Fundamentals, Evolving Technologies, and Emerging Applications. CRC Press (2012)
8. Langlois, J.H., Roggman, L.A.: Attractive faces are only average. Psychol. Sci. 1(2), 115121 (1990)
9. Liu, L., Sheng, Y., Zhang, G., Ugail, H.: Graph cut based mesh segmentation using feature points and geodesic distance. In: Cyberworlds 2015, Gotland, Sweden, pp. 115–120 (2015)
10. Loth, S.R., Iscan, M.Y.: Sex Determination, Encyclopedia of Forensic Sciences, vol. 1. Academic Press, San Diego (2000)
11. Ugail, H., Al-dahoud, A.: Is gender encoded in the smile? A computational framework for the analysis of the smile driven dynamic face for gender recognition. Vis. Comput. 34(9), 12431254 (2018)
12. Yap, M.H., Ugail, H., Zwiggelaar, R.: Intensity score for facial actions detection in near-frontal-view face sequences. Comput. Commun. Eng. 6, 819–824 (2013)
13. Yap, M.H., Ugail, H., Zwiggelaar, R.: Facial analysis for real-time application: a review in visual cues detection techniques. J. Commun. Comput. 9, 1231–1241 (2013)
14. Yap, M.H., Ugail, H., Zwiggelaar, R.: Facial behavioural analysis: a case study in deception detection. Br. J. Appl. Sci. Technol. 4(10), 1485–1496 (2014)

Chapter 5
The Biometric Characteristics of a Smile

Abstract Facial expressions have been studied looking for its diagnostic capabilities in mental health and clues for longevity, gender and other such personality traits. The use of facial expressions, especially the expression of smile, as a biometric has not been looked into great detail. However, research shows that a person can be identified from their behavioural traits including their emotional expressions. In this Chapter, we discuss a novel computational biometric model which can be derived from the smile expression. We discuss how the temporal components of a smile can be utilised to show that similarities in the smile exist for an individual and it can be enabled to create a tool which can be utilised as a biometric.

Keywords Smile biometrics · Smile dynamics · Smile intervals · Clustering

5.1 Introduction

A biometric method can be an automated mechanism through which someone's physiological or behavioural characteristics can be used to uniquely identify that person [4]. Human physiological or behavioural characteristics show high levels of complexity. Common methods of biometric identification follow two major routes: psychological and behavioural. Physiological characteristics of biometric-based technologies include models such as the face, fingerprints, hand geometry, iris, retina, ear and voice. Behavioural characteristics include models such as gait signature and keystroke dynamics. Both physiological and behavioural characteristics can be categorised based on human involvement which can be voluntary or involuntary. A voluntary model, which requires user involvement to perform the authentication process, includes iris, hand and finger. The involuntary model, which does not require user involvement to perform the authentication process, includes face and voice. All involuntary models can be executed passively without any explicit action or participation on the part of the user.

Here, we discuss a novel behavioural biometric model which is the facial expression biometric [1, 3]. A facial expression biometric can be identified as the authentication process done through examining the way the person behaves and expresses emotions, e.g. [6]. Research on facial expression biometrics is very limited and there are a couple of psychological and computational studies that try to identify the facial expression biometrics and use it as a biometric technique. Work in this area is mainly carried out through the use of EMG, or other psychological studies and through the use of visual computing techniques. And interestingly, all the work has been undertaken via the use of smile expression as it is considered to be one of the most sophisticated facial expressions.

Cohn et al. [2] undertook a study to find out the stability of smile facial expressions. Using EMG to measure the zygomatic major muscle through AU12 they show that the AU12 has a stable variation through a two-year study for the same individual. Furthermore, they report that recognising an individual based on their EMG reading through the identification of facial action units is above chance. Similarly, Schmidt and Cohn [5] studied the stability of the smile expressions through time. In their research, they studied AUs 6 and 12 to analyse 195 smiles from 95 individuals. Using automated facial software analysis and EMG to analyse spontaneous smiles, their study covered two sessions with one year interval. As a result, they conclude, facial EMG provides evidence to support the stability of the smile expression over time, for the same individual.

It is, therefore, an interesting proposal to see the possibility of using computational facial expressions as a biometric—in particular, to show how the smile expression can be unique. Furthermore, our approach tries to identify the dynamic aspect of the smile expression where we study the stabilities of the different smiles for the same subject and the similarity aspects of such smiles. We apply a set of experiments to try identify a potential smile biometric. First, we use a machine learning technique to identify the possibility of using the smile expression as a biometric. Second, we propose a novel algorithm that uses only the dynamic features of the smile and tries to identify it as a biometric. The benefit of this research is that it can identify another dimension of human emotions which can have unique applications.

5.2   Proposed Method

The framework we have developed utilises the dynamic components of the smile. Figure 5.1 illustrates the computational framework that can be adopted for this purpose. Face detection, face processing and landmark detection are undertaken through the Viola Jones algorithm and the CHEHRA model, as discussed in the previous chapters.

5.2.1   Smile Intervals

The smile intervals represent the time at which the smile starts to reach the peak (the onset) and stay there (the peak) and then return to the neutral expression (the

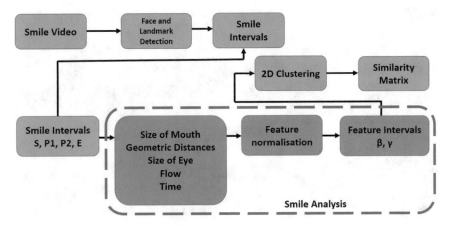

Fig. 5.1 Block diagram showing the key components of the computational framework for facial emotion recognition

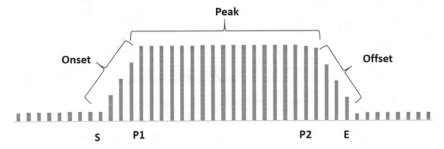

Fig. 5.2 Changes in the mouth size through the smile expression with various smile intervals identified

offset). Computing the smile intervals is composed of three main parts. They are the identification of the start of the smile S, computing the peak of the smile which is identified by the interval $P1$ and $P2$ and identification of the smile returning to the neutral expression defined as E. Figure 5.2 illustrates this further.

Using the landmarks we detect, we represent the mouth and the eye area using triangular features as shown in Fig. 5.3. These triangular features are then summed up to get the total mouth size m_1, the eye sizes e_1 and e_2. Table 5.1 shows details of the features utilised for the smile analysis along with their descriptions and how they can be computed.

To detect the peak interval $(P1, P2)$, first, we compute the "peak average" which is done using the mouth area m_1, whereby we compute the average from the frame S to the end of the smile. Computing the peak average creates a imaginary threshold line that identifies the peak of the smile interval. To identify $(P1, P2)$ we scan the value of m_1 through the smile expression from the start to the end of the smile and compare it to the peak average. P_1 is identified by checking the mouth size at each

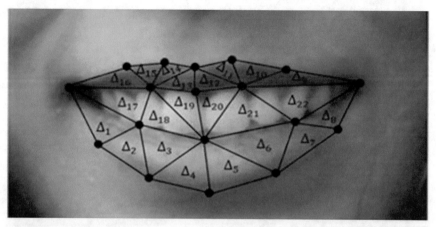

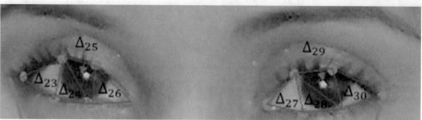

Fig. 5.3 Representation of the mouth and eyes using triangular features

Table 5.1 Description of the dynamic smile features and how they can be computed

Feature	Description	How the feature is computed
d_1	Distance between mouth corners	Distance (P_{32}, P_{38})
m_1	Total mouth area	$\sum_{i=1}^{22} \Delta_i$
LL_1	Lower lip area	$\sum_{i=1}^{8} \Delta_i$
UL_1	Upper lip area	$\sum_{i=9}^{16} \Delta_i$
e_1	Left eye area	$\sum_{i=23}^{26} \Delta_i$
e_e	Right eye area	$\sum_{i=27}^{30} \Delta_i$

frame to the peak average which is identified from P_1 from the start of the smile to the end of it. P_2 is identified by checking the mouth size at each frame to the peak average from the end of the smile to P_1. To identify the end of the smile E, we compare the changes in d_1 to the neutral frame d_N, from P_2 to the end of the smile.

5.2.2 Smile Analysis

After identifying the smile intervals, we can measure a set of features to infer the smile dynamics. To do this, we can analyse the smile within the time intervals we identified. The analysis of the smile include identifying various features such as,

- distance changes between corners of the mouth,
- changes in size of the mouth,
- changes in the size of the upper and lower lips,
- changes in the eye area,
- optical flow values,
- and the order of smile feature activation.

The selection of these features is based on the landmark detection model and the dynamic characteristics these features have through smile expression. After computing each feature, we normalise it by dividing with the triangle formed by the tip of the nose and the outer corners of the eyes. The normalisation process unifies the finding for each subject and eliminates distance factors relative to the camera, for example.

Once we compute the dynamic features, we can use various machine learning techniques to analyse the dynamic flow of each feature. These computations all may contain small errors. Additionally, since we have more than one smile for each subject, there may be small variations in computations of the dynamics of the smile. To compensate for such potential errors, we can compute the feature intervals represented by β and γ—where β represents the high features interval and γ represents the low interval of the features. Combining β and γ creates a imaginary cluster that represents the average of the smile dynamics for a specific subject.

5.2.3 Computing Similarities

For classification, we can use a 2D-clustering approach where we can use both β and γ to identify the boundary for each feature cluster related to the specific subject. These clusters represent different features in a 2D space with a boundary set through determination of β and γ. To compute the smile similarity though utilisation of the proposed features, we can compare the dynamics of each feature. We can check if it is within the boundary β and γ or not. It is then represented as a score and can be identified with a specific subject. Summing up all the smile scores for different features represents the smile similarity score. For visualisation purposes, Fig. 5.4 shows two clusters for $D1$ and $M1$ for one subject in the MUG dataset.

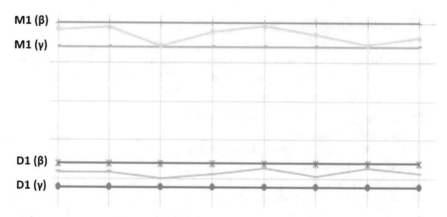

Fig. 5.4 Visualisation of the 2D clustering

5.3   Results

The analysis, in this case, was done on the MUG dataset where we used 20 subjects. Each subject had an average of three smiles. To test our smile interval detection, we compared our proposed method to the manual set of smile intervals which we added to the MUG dataset. Figure 5.5 shows the proposed method of detecting intervals $S - P, P1 - P, P2 - P, E - P$ versus the manual time intervals $S, P1, P2, E$ for 10 subjects. Our algorithm has a 94% match with the manually coded smile intervals.

To test the proposed dynamic features, we can first examine how many similarities each feature contains. To find such similarities, we can use the 2D-clustering

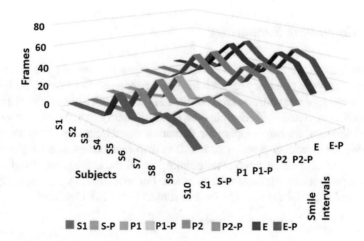

Fig. 5.5 Identification of the smile intervals—manual versus proposed

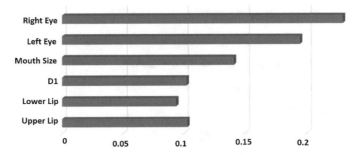

Fig. 5.6 Similarity of the dynamic smile features at the onset interval

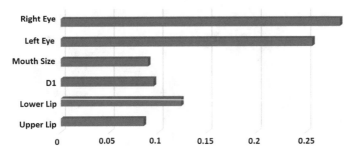

Fig. 5.7 Similarity of the dynamic smile features at the peak interval

technique to see how much each feature is similar or unique for all the smiles in the MUG dataset. In the first experiment, we compute similarity for each time interval S, $P1 - P2$ and E. Figure 5.6 shows the similarity percentages for the dynamic facial features at the start of the smile.

Figures 5.6, 5.7 and 5.8 show the similarity graphs for the facial features at the onset, peak and offset. It can be observed the uniqueness of each facial feature in the peak interval. As an example, as shown, the results around lower lip shows 10% similarity which implies that three smiles have similar lower lip dynamics in the peak period.

In another experiment, we compute the average flow similarity for each smile interval. Figure 5.9 shows the average similarity by considering the total flow in the facial features of each subject, normalised for each smile interval.

Finally, we compute time features, where we compute the similarity for the time intervals between the subjects in the MUG dataset. Figure 5.10 shows the smile interval similarity. Using the total time as a signature shows that 8% of the subjects use the same timing.

Since we use the flow value for computing the total displacement, as shown in Fig. 5.8, we can also use it to compute the order of smile activation. After checking the order of activation, we can check for any variations within the same subject in terms of the order they use facial features to formulate the smile. Additionally, the results show that the mouth region has the highest flow value, cheeks the second and

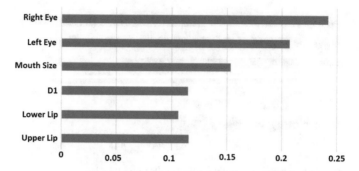

Fig. 5.8 Similarity of the dynamic smile features at the offset interval

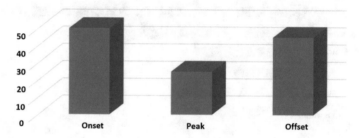

Fig. 5.9 Visualisation of the flow similarity within smile intervals

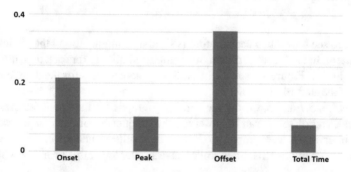

Fig. 5.10 Visualisation of the flow similarity across various time intervals

eyes the lowest. These findings diverted our attention to see which parts of the face has more weight in smile expression (left or right). Using the proposed method, the result shows partial stability of using the left or right side of the face for the same subject. The results show that 60% of the subjects in the MUG dataset use their left side of the face and the other 40% use the right side. Due to the weakness of this feature, we eliminate it from our computation of the similarity matrix.

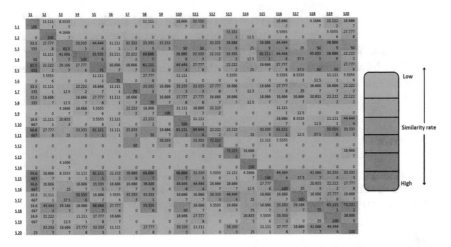

Fig. 5.11 A heat map depicting the similarity in the dynamics of the smile expression showing its uniqueness

For validation of our approach, we compare all the feature clusters for all the subjects in the MUG dataset. We use β and γ to identify the overall similarity for each subject in the dataset. Figure 5.11 shows a similarity heat map for the subjects whereby we have selected the highest scores for each subject.

5.4   Conclusion

This chapter presents a novel framework that can be utilised to study if the dynamic features of the smile can be used as a biometric. The approach here is to study only the dynamic characteristics of the facial features through the smile expression. First, we can identify the smile intervals which represent various parts of the smile in terms of time, such as the start point, the peak points and the end of the smile. Using time intervals, we can focus on analysing various dynamic features which include, upper and lower lips, mouth area, eyes, the distance between the mouth corners, the optical flow on the face and other temporal characteristics. The proposed algorithm identifies the smile intervals showing a 94% correct classification against the manually coded intervals in the MUG dataset.

For classification, we can make use of the feature intervals which are taken as a 2D cluster that captures different variations of features through the smile expression. Feature intervals can be computed using the average and the standard deviation for a different smile for the same subject and can be used to measure the degree of smile similarities among an individual.

The results firstly indicate that humans have stability in their smiles which include mouth size, smile time and other features through the smile expression. Secondly, it can be observed that there exist a great degree of individual similarity in the smiles among the subjects in the MUG dataset. Thirdly, the results indicate that the smile dynamics can be used to uniquely identify the individuals. Finally, we believe the study of additional features in the smile such as the order of activation of different components of the smile would be interesting.

References

1. Andre, M.G.C., Nummenmaa, L.: Facial expression recognition in peripheral versus central vision: role of the eyes and the mouth. Psychol. Res. **78**(2), 180195 (2014)
2. Cohn, J., Schmidt, K., Gross, R., Ekman, P.: Individual differences in facial expression: stability over time, relation to self-reported emotion, and ability to inform person identification. In: Proceedings of the International Conference on Multimodal User Interfaces (ICMI 2002), pp. 491–496 (2002)
3. Ding, C., Tao, D.: Robust face recognition via multimodal deep face representation. IEEE Trans. Multimed. **17**(11), 20492058 (2015)
4. Elmahmudi, A., Ugail, H.: Experiments on deep face recognition using partial faces. In: Cyberworlds 2018. IEEE (2018)
5. Schmidt, K., Cohn, J.: Dynamics of facial expression: normative characteristics and individual differences. In: IEEE International Conference on Multimedia and Expo, pp. 728–731 (2001)
6. Ugail, H.: Secrets of a smile? Your gender and perhaps your biometric identity. Biom. Technol Today **2018**(6), 57 (2018)

Chapter 6
Conclusions

Abstract If the face is a window to the soul, then the smile is the light that reflects from the soul. The face conveys much information about a person, be it the identity, gender, feelings or even the thought process, e.g. [10, 13–15]. Since the smile is one of the most complex facial expressions, it is of no surprise that it contains much personality traits and other information about the individual.

Rigorous scientific studies on facial emotional expressions, including the smile expression, dates back to the 1800s. Duchenne postulated that the face is some kind of a map which can be utilised to understand the mental states of a human [2]. Darwin carried out systemic experiments on facial emotions in cross-cultural settings [1]. Then, in the 1970s, computer scientists took up the challenge of studying facial expressions using the power of numerical computations. A flurry of research activity continued with researchers interested in developing methods and techniques for automated facial analysis, e.g., [3–9]. This resulted in algorithms and systems for facial expression analysis and annotation. With such advances in automated facial expression analysis, many discoveries have been made including many personality and behavioural traits as well as clinical diagnoses.

As far as computational facial expression analysis is concerned, the current state of the art in the algorithmic developments include three significant stages. First, the face is detected and processed—whereby faces are automatically identified from a video and a given set of landmarks are automatically labelled on the face. The second step is feature extraction from the face. Such features can be pure geometry based or appearance based, or it can be features which machine learning algorithms may extract. The final step in a computational facial emotion analysis system is the classification of a given expression or set of expressions or emotions.

This book has been concerned with computational techniques for human smile analysis. First, we discussed a framework to enhance the detection of facial features as it plays a significant role in detecting and analysing the smile expression. Second, we discussed how a smile expression can be tested for its genuineness. More specifically, we show how the smile is distributed across the face over the individual features of the

© The Author(s), under exclusive licence to Springer Nature Switzerland AG 2019
H. Ugail and A. A. A. Aldahoud, *Computational Techniques for Human Smile Analysis*,
SpringerBriefs in Computer Science, https://doi.org/10.1007/978-3-030-15381-6_6

face. Third, we discussed how the smile expression can be utilised as a tool for gender classification. More specifically, it is shown how the spatio-temporal properties of the smile can be used to infer gender. Finally, we have presented a computational framework through which one can study the biometric traits of a smile, showing how the dynamics of the smile expression is stable at the level of an individual.

In terms of applications, in this book, we have discussed how genuine and posed smiles can be distinguished, how gender can be inferred from the smile and how it is possible to utilise the smile expression as a form of biometric. These applications are among many that a smile can serve. Other applications include analysis of pain, diagnosis of depression, detection of deception, expression transfer for human-computer interaction and entertainment purposes.

When it comes to the development of tools for automated computational facial expression analysis, including tools for the smile expression analysis, there are still a great many challenges to overcome. Efficient and robust real-time systems for face detection, facial landmark identification, representation, feature extraction and classification are needed. Current systems are often limited to the laboratory or to indoor scenarios. There is a great need for more robust and versatile systems to be developed so as the full breadth of applications can be realised. The ability for computational schemes to develop complex models is limited at this stage due to the lack of availability of sufficient datasets. In addition to this, much of the existing facial emotion classification schemes and taxonomies have been developed using observer based human judgements. For example, a great deal of smile expression analysis work is done using FACS coded videos. The development of sophisticated unsupervised machine learning techniques in this field will greatly assist in helping it to advance further.

The smile is the light that reflects from the soul. To learn more about the soul, may we study the smile expression in much greater detail.

References

1. Darwin, C.: The Expression of the Emotions in Man and Animals. John Murray, London (1872)
2. Duchenne, G.-B., Cuthbertson, A.R.: The Mechanism of Human Facial Expression. Cambridge University Press (1990)
3. Ekman, P.: An argument for basic emotions. Cogn. Emot. **6**(3/4), 169–200 (1992)
4. Ekman, P.: Telling Lies. Norton, New York (2009)
5. Fridlund, A.J.: Human Facial Expression: An Evolutionary View. Academic Press, New York (1994)
6. Han, X., Ugail, H., Palmer, I.: Gender classification based on 3D face geometry features using SVM. In: Cyberworlds 2009, Bradford, UK (2009)
7. Pantic, M., Rothkrantz, L.: Expert system for automatic analysis of facial expression. Image Vis. Comput. **18**, 881–905 (2000)
8. Schmidt, K.L., Ambadar, Z., Cohn, J.F., Reed, L.I.: Movement differences between deliberate and spontaneous facial expressions: Zygomaticus major action in smiling. J. Nonverbal Behav. **30**, 37–52 (2006)

9. Ugail, H., Al-dahoud, A.: Is gender encoded in the smile? A computational framework for the analysis of the smile driven dynamic face for gender recognition. Vis. Comput. **34**(9), 12431254 (2018)
10. Ugail, H.: Secrets of a smile? Your gender and perhaps your biometric identity. Biom. Technol. Today **2018**(6), 57 (2018)
11. Yap, M.H., Ugail, H., Zwiggelaar, R., Rajoub, B., Doherty, V., Appleyard, S., Huddy. G.: A short review of methods for face detection and multifractal analysis. In: Cyberworlds 2009, Bradford, UK (2009)
12. Yap, M.H., Ugail, H., Zwiggelaar, R., Rajoub, B.: Facial image processing for facial analysis. In: IEEE International Carnahan Conference on Security Technology (ICCST 2010), California, USA, San Jose (2010)
13. Yap, M.H., Ugail, H., Zwiggelaar, R.: Intensity score for facial actions detection in near-frontal-view face sequences. Comput. Commun. Eng. **6**, 819–824 (2013)
14. Yap, M.H., Ugail, H., Zwiggelaar, R.: Facial analysis for real-time application: a review in visual cues detection techniques. J. Commun. Comput. **9**, 1231–1241 (2013)
15. Yap, M.H., Ugail, H., Zwiggelaar, R.: Facial behavioural analysis: a case study in deception detection. Br. J. Appl. Sci. Technol. **4**(10), 1485–1496 (2014)

Index

© The Author(s), under exclusive licence to Springer Nature Switzerland AG 2019   61
H. Ugail and A. A. A. Aldahoud, *Computational Techniques for Human Smile Analysis*,
SpringerBriefs in Computer Science, https://doi.org/10.1007/978-3-030-15381-6

Printed in the United States
By Bookmasters